T0228206

The Art of the Soluble

First published in 1967, *The Art of the Soluble* presents collection of essays giving the views of the author on creativity and originality in science and on the logical connections between creative and critical thought. It is also a pioneering study of the ethology of the scientists – of the anatomy of scientific behaviour. Is it true that scientists are detached or dispassionate observers of Nature? What underlies the scientist's deep concern over matters of priority? How did a class distinction grow up between pure and applied science? By what criteria do scientists value their own and their colleagues work? Some of the answers grow out of the author's four critical studies of Teilhard de Chardin, Arthur Koestler, D'Arcy Thompson and Herbert Spencer and the book as whole is knit together by a major essay *Hypothesis and Imagination*, on the nature of scientific reasoning. P. B. Medawar, who won the Nobel Prize for Medicine in 1960, did not see science as a book-keeping of Nature but, on the contrary, as the greatest of human adventures. This book will be an essential read for scholars and researchers of philosophy of Science, natural science, and philosophy in general.

The Art of the Soluble

by P.B. Medawar

Routledge
Taylor & Francis Group

First published in 1967
By Methuen & Co.

This edition first published in 2021 by Routledge
2 Park Square, Milton Park, Abingdon, Oxon, OX14 4RN
and by Routledge
605 Third Avenue, New York, NY 10017

Routledge is an imprint of the Taylor & Francis Group, an informa business

Publisher's Note
The publisher has gone to great lengths to ensure the quality of this reprint but points out that some imperfections in the original copies may be apparent.

Disclaimer
The publisher has made every effort to trace copyright holders and welcomes correspondence from those they have been unable to contact.

A Library of Congress record exists under LCCN: 67083875

ISBN 13: 978-1-032-11681-5 (hbk)
ISBN 13: 978-1-003-22103-6 (ebk)
ISBN 13: 978-1-032-11685-3 (pbk)

DOI: 10.4324/9781003221036

THE ART OF
THE SOLUBLE

P. B. MEDAWAR, F.R.S.

METHUEN & CO LTD
11 New Fetter Lane, London EC4

This collection first published 1967 by
Methuen & Co. Ltd., 11 New Fetter Lane, London EC4
© *1967, 1958, 1961, 1963, 1964, 1965 P. B. Medawar*
Printed and bound in Great Britain by
Cox & Wyman Ltd, Fakenham, Norfolk

CONTENTS

TO JEAN

INTRODUCTION

The title of this collection comes from the review on pp. 85–96 of Arthur Koestler's 'The Act of Creation' *(New Statesman, 19 June 1964)*:

> No scientist is admired for failing in the attempt to solve problems that lie beyond his competence. The most he can hope for is the kindly contempt earned by the Utopian politician. If politics is the art of the possible, research is surely the art of the soluble. Both are immensely practical-minded affairs.

Good scientists study the most important problems they think they can solve. It is, after all, their professional business to solve problems, not merely to grapple with them. The spectacle of a scientist locked in combat with the forces of ignorance is not an inspiring one if, in the outcome, the scientist is routed. That is why some of the most important biological problems have not yet appeared on the agenda of practical research.

In so far as these essays have a central or recurrent theme, it lies in the attempt to answer two questions: what sort of a person is a scientist, and what kind of act of reasoning leads to scientific discovery and the enlargement of the understanding? I seldom put these questions directly, and answer them only very incompletely and bit by bit. Two articles deal with the work and thought of particular people. 'Herbert Spencer and the Law of General Evolution' was the Spencer Lecture delivered at Oxford in 1963 and printed in *Encounter* in September 1963. To prepare myself for the occasion I made up my mind to read deeply in Spencer himself: a grave decision, as those who know his works will testify, but justified in the outcome, for Spencer does not deserve the neglect he has fallen into; many of his ideas are now being served up anew with only a patronizing nod by way of acknowledgement, and in a form of which he himself would very much have disapproved. The Spencer Lecture begins with an account of the logical weaknesses of his or any other theory of general evolution – of any theory which declares that evolution is the fundamental stratagem of Nature, is 'a general condition to which . . . all systems must bow . . . a light illuminating all facts, a curve all lines must follow', to put the matter in words (Teilhard's) which would not have commended themselves to Spencer. Yet the theory of Evolution in its ordinary technical sense is still one of the liveliest topics in biology, and the next great advance we may look forward to is a new theory of genetic variation, propounded by molecular biologists. For the reasons explained in my Presidential Address to Section D of the British Association ('A Biological Retrospect', published in *Nature*, 25 September 1965), it is impossible to predict what this advance will be, but it will clearly be something in the nature of a sentential calculus of genetic messages – a natural development of the glossary that is now being compiled of single words.

The latter part of the Spencer Lecture deals with certain abstract formal similarities between natural transformations of energy, information theory, probability, and the idea of order; or, alternatively, between entropy, randomness and nonsense.

The second article that deals with the work of a particular man is 'D'Arcy Thompson and *Growth and Form*'. It first appeared as a postscript to his daughter Ruth D'Arcy Thompson's biography (Oxford University Press, 1958) and was reprinted by *Perspectives* in its winter issue of 1962. Because D'Arcy Thompson was a classicist and a mathematician as well as a natural historian, it may seem specially significant that he believed, as I also believe,

. . . not merely that the physical sciences and mathematics offer us the only pathway that leads to an understanding of animate nature, but also that the true beauty of nature will be revealed only when that understanding has been achieved. To us nowadays it seems obvious that the picture we form in our minds of Nature will be the more beautiful for being brightly lit. To many of D'Arcy's contemporaries it must have seemed strange and even perverse that he should have combined a physico-mathematical analysis of Nature with, at all times, a most intense consciousness of its wonder and beauty; for at that time there still persisted the superstition that what is beautiful and moving in Nature is its mystery and its un-revealed designs. D'Arcy did away for all time with this Gothick nonsense: a clear bright light shines about the pages of *Growth and Form*, a most resolute determination to unmake mysteries.

This last sentence is my cue to mention the highly critical review (in *Mind*, January 1961) of Pierre Teilhard de Chardin's *The Phenomenon of Man*. A good deal of Teilhard is nonsense, but on further reflection I see it as a dotty, euphoristic kind of nonsense, very greatly preferable to solemn long-faced germanic nonsense. There is no real harm in it. But what, I wonder, was the origin of the really mischievous belief that obscurity makes a *prima facie* case for profundity? – the origin, I mean, of the syllogism that runs *Profound reasoning is difficult to understand: this work is difficult to understand: therefore this work is profound.*

In the seventeenth and eighteenth centuries philosophic writing was marked by its clarity. John Locke is easier to follow than John Donne. Descartes assured even his most diffident readers[1] that there was nothing in his writings 'which they are not capable of completely understanding, if they take the trouble to examine them', and the ideas of clarity and of distinctness of vision occur as often in his writings as the idea of light, the *lumen siccum*, in Bacon's.

Hume and Berkeley, Reid and Stewart, had definite opinions to express and took care to make them fully understood; but with Kant things changed. Kant was a very profound philosopher, in some ways the greatest there has ever been. One of the reasons why he is hard to understand is because the problems he tried to solve are intrinsically of the utmost difficulty. We marvel at and

[1] In a letter to the translator (into French) of his *Principles of Philosophy*.

revere his struggle for clarity. Unfortunately, his struggle was sometimes unsuccessful. Kant became notorious for his obscurity. Dugald Stewart, wrestling with the Sage in Latin translation, spoke of his 'utter inability to unriddle the author's meaning ... 'I have always been forced to abandon the undertaking in despair'. Seen through Peacock's eyes, the young Shelley delighted in

... the sublime Kant, who delivers his oracles in language none but the initiated can comprehend,

and Coleridge

... plunged into the central opacity of Kantian metaphysics, and lay *perdu* for several years in transcendental darkness. ...

Kant was just the man for the Romantics.

The harm Kant unwittingly did to philosophy was to make obscurity seem respectable. From then on, any petty metaphysician could hope to be given credit for profundity if what he said was almost impossible to follow.[1] There grew up a new style of philosophic writing of which F. H. Bradley was the greatest English master. It seems to have affected many of its readers like a drug, and the intense resentment aroused by the work of the modern Oxford philosophers can be thought of as part of a withdrawal syndrome, for in England metaphysics in the Bradleian style has almost completely disappeared. Of course we all need to be *bunkrapt* from time to time (Paul Jennings's word), but not by works that profess to be philosophy.

'Two Conceptions of Science' was the Henry Tizard Memorial Lecture, delivered at Westminster School in 1965 and reprinted in *Encounter* in August. The two conceptions are, roughly speaking, the romantic and the rational, or the poetic and analytical, the one speaking for imaginative insight and the other for the evidence of the senses, one finding in scientific research its own reward, the other calling for a valuation in the currency of practical use. None

---

[1] Kant had to face this charge himself: 'I am often accused of obscurity, perhaps even of deliberate vagueness in my philosophical discourse to give it the air of deep insight' (Preface, *The Metaphysics of Morals*).

of these distinctions is really satisfactory, for the two sets of opinions they try to embody do not hang together logically or rationally in themselves; rather they are complexes of opinion that tend to go with certain temperaments, much as Tory and Labour or Republican and Democrat stand for casts of thought and not merely for casts of vote. Nevertheless I think I am right in saying that in the Romantic conception

> . . . truth takes shape in the mind of the observer: it is his imaginative grasp of *what might be true* that provides the incentive for finding out, so far as he can, what *is* true. Every advance in science is therefore the outcome of a speculative adventure, an excursion into the unknown. According to the opposite view, truth resides in nature and is to be got at only through the evidence of the senses: apprehension leads by a direct pathway to comprehension, and the scientist's task is essentially one of *discernment*.

It is the so-called 'hypothetico-deductive' scheme of thought that makes sense of and reconciles these two apparently contradictory sets of opinions. 'Hypothesis and Imagination' is a commentary on (and the beginnings of a history of) the hypothetico-deductive system; it is an expanded and annotated version of an article with the same title that appeared in the *Times Literary Supplement* of 25 October 1963, and it was written for a forthcoming volume in *The Library of Living Philosophers* as a tribute to our foremost methodologist of science, Karl Popper. So far as the historical matter goes, I have been concerned mainly with English and Scottish philosophical thought, and with other philosophic traditions only in so far as they touch upon or mingle with it.

To go back to 'Two Conceptions of Science': I am not now at all satisfied with the argument that occupies its latter half, which deals with why, in this country in particular, a class distinction has grown up around the difference between 'pure' and 'applied' science. My suggestion ran as follows: in this country, poetic invention is the paradigm of pure creative activity. The Romantics shunned 'applied' poetry – poetry for the occasion, or poetry upon a given theme – because they felt that artistic creation should be a natural and spontaneous upwelling of ideas. Correspondingly, the highest form of science must be that which is spontaneously proffered by the creative imagination, not something wrung from us by the pressure of necessity.

This interpretation struck me as plausible not only because the Romantics championed it, but also because our reverence for pure science in the modern sense (see below) does in fact date from the first half of the nineteenth century. In the seventeenth and eighteenth centuries the idea of science pursued for its own sake was regarded as frivolous or even comic: that is why Francis Bacon and Thomas Sprat had to beg for interpretative research, for *experiments of light*. Sir Nicholas Gimcrack, Otway's *Virtuoso*, full of the 'purest' scientific enterprises, is an egregious ass. Goldsmith had to plead the usefulness of his *History of the Earth and Animated Nature* because the opposite of useful was not pure but – *idle*. Swift's ridiculous Laputans were Fellows of the Royal Society. It went with this cast of mind that hypotheses were thought a perilous indulgence (see pp. 138–142), and the Creative Imagination was looked upon with disfavour.

But Coleridge was not against the Useful Arts. Agriculture, commerce and manufactures, he tells us,[1] 'are now considered scientifically', and are indeed founded on the sciences:

> It is not, surely, in the country of ARKWRIGHT, that the Philosophy of Commerce can be thought independent of Mechanics: and where DAVY has delivered Lectures on Agriculture, it would be folly to say that the most Philosophic views of Chemistry were not conducive to the making our valleys laugh with corn.

Coleridge certainly thought the Useful Arts less noble than the Pure Sciences – but because of their fallibility, not because of their usefulness. Coleridge used 'pure' in its older and original meaning. The Pure Sciences are those of which the axioms are derived, not

[1] Coleridge *On Method*, 3rd Ed. (London 1849).

from experience, but by intuition or revelation: the 'Pure Sciences . . . represent pure acts of the Mind', Coleridge says; '. . . at the head of all Pure Science stands *Theology*, of which the great fountain is Revelation'. Pure Sciences are best not because they lack practical application but because the revealed or intuitively certain axioms out of which they grow are not subject to error.

Over the past hundred years or so the word 'pure' in this context has undergone a slow revolution of meaning which has allowed its lesser connotation to usurp the greater; 'pure science' now means that which is done without regard to or interest in application or use, and we speak of 'pure physics' or 'pure biological research' in a sense that differs in an important way from Coleridge's; yet the idea of the superiority of pure over applied science has remained, and with it the dire equation *Useless = Good*.

Many other influences were of course at work. Perhaps the most important of all – and the best understood, in spite of my failure to have given it due weight – was the conscious and deliberate perpetuation by our public schools of the Platonic conception of activities that did or did not become a gentleman.[1] It was put admirably by John Gillies in the introduction to his translation of Aristotle's *Ethics and Politics* (London, 1797). Aristotle's experimentations, he tells us, were 'confined to catching Nature in the fact, without attempting, after the modern fashion, to put her to the torture'; for philosophers

> . . . ranked with the first class of citizens; and, as such, were not lightly to be subjected to unwholesome or disgusting employments. To bend over a furnace, inhaling noxious steams; to torture animals, or to touch dead bodies, appeared to them operations . . . unsuitable to their dignity. For such discoveries as the heating and mixing of bodies offer to inquisitive curiosity, the naturalists of Greece trusted to slaves and mercenary

[1] See Mr J. C. Dancy's presidential address to the Educational Section of the British Association in 1965, in *The Advancement of Science*, Vol. 22.

mechanics, whose poverty or avarice tempted them to work in
metals or minerals. . . . The work shops of tradesmen then
revealed those mysteries which are now sought for in colleges
and laboratories . . .

These canons of gentility were still in force and gloried in during
my own schooldays. They are quite largely responsible for the
position Great Britain holds in the world today. It is not envy or
malice, as so many people think, but utter despair that has per-
suaded many educational reformers to recommend the abolition
of the English public schools.

'Darwin's Illness' was published in the *New Statesman* of 3
April 1964. When Professor Saul Adler hit upon the idea, now
taken very seriously by many clinicians, that Darwin had been the
lifelong victim of South American trypanosomiasis, a chronic and
debilitating disease, I thought it would be interesting to study
anew the opinions of those who had confidently declared for a
purely psychogenic interpretation of his ailments, though before
I began I had no idea of what treasures of nonsense I should find in
the psychoanalytical literature. Many good people contend that,
while the theoretical foundations of psycho-analysis leave much
to be desired, yet, as a method of treatment, it demonstrably
works: people do sometimes, perhaps often, get better as a result of
psycho-analytic treatment – we have their own word for it, and
who should know better?

That people improve *under* psycho-analytic treatment can
hardly be a matter of dispute, but I know no evidence which
shows that they get better as a consequence of psycho-analytic
treatment as such. Many people seem to have no clear understanding of what constitutes valid evidence that a medical treatment is efficacious. Even those who pride themselves on their good
judgement (and deplore the lack of it in others) are almost completely gullible when it comes to weighing up the value of any
treatment they may have received themselves. If a person *(a)* is
poorly, *(b)* receives treatment intended to make him better, and
(c) gets better, then no power of reasoning known to medical

science can convince him that it may not have been the treatment that restored his health. Psychological treatments are specially difficult to evaluate. For example: a young man full of anxieties and worries may seek treatment from a psycho-analyst, and after eighteen months' or two years' treatment finds himself much improved. Was psycho-analytic treatment responsible for the cure? One cannot give a confident answer unless one has reasonable grounds for thinking

(a) that the patient would not have got better anyway;

(b) that a treatment based on quite different or even incompatible theoretical principles, e.g. the theories of a rival school of psycho-therapists, would not have been equally effective; and

(c) that the cure was not a by-product of the treatment. The assurance of a regular sympathetic hearing, the feeling that somebody is taking his condition seriously, the discovery that others are in the same predicament, the comfort of learning that his condition is explicable (which does not depend on the explanation's being the right one) – these factors are common to most forms of psychological treatment, and the good they do must not be credited to any one of them in particular. At present there is no convincing evidence that psycho-analytic treatment as such is efficacious, and unless strenuous efforts are made to seek it the entire scheme of treatment will degenerate into a therapeutic pastime for an age of leisure.

It occurred to me, when preparing an address to Reed College on the theme 'Science and the Sanctity of Life',[1] that nowadays we all give too much thought to the material blessings or evils that science has brought with it, and too little to its power to liberate us from the confinements of ignorance and superstition.

The greatest liberation of thought achieved by the scientific revolution was to have given human beings a sense of a future in this world. The idea that the world has a virtually indeterminate future is a comparatively new one. Much of the philosophic

---

[1] Printed in *Encounter*, November 1966.

speculation of three hundred years ago was clouded over by the thought that the world had run its course and was coming shortly to an end.[1] 'I was borne in the last age of the world', said Donne, giving it as the 'ordinarily received' opinion that the world might run thrice two thousand years between its creation and the Second Coming; and according to Archbishop Ussher's chronology more than five and a half of those six thousand years had gone by already.

No empirical evidence challenged this dark opinion. There were no new worlds to conquer, for the world was known to be spherical and therefore finite; certainly it was not all known, but the full extent of what was *not* known was known. Outer space did not put into people's minds then, as it does into ours now, the idea of a tremendous endeavour only just beginning. Moreover, life itself seemed changeless. The world a man saw about him in adult life was much the same as it had been in his own childhood, and there was no reason to think it would change in his own or his children's lifetime. We need not wonder that the promise of the next world was held up to believers as an inducement to put up with the incompleteness and inner pointlessness of this one: the present world was only a staging post on the way to better things. There was a certain awful topicality about Thomas Burnet's tremendous description of the world in ruins at the end of its long journey from 'a dark chaos to a bright star', for the end of the world might indeed be near at hand. (Burnet was not very precise about dates, but he seemed to be thinking in terms of a standard deviation of about a hundred years.) And Thomas Browne warned us against the folly and extravagance of raising monuments and tombs intended to last many centuries; we are living in the Setting Part of Time, he told us; *the Great Mutations of the World are acted: it is too late to be ambitious.* But science has now made it the ordinarily received opinion that the world has a future extending far beyond the most distant frontiers of what can be rationally imagined, and that is perhaps why, in spite of all

[1] See *The Discovery of Time* by J. Goodfield and S. Toulmin. Macmillan, 1965.

his faults, scientists still incline to think Francis Bacon their first and greatest spokesman: we may yet build a New Atlantis.

The point is that when Thomas Burnet exhorted us to become Adventurers for Another World, *he* meant the next world; but we mean this one.

I am grateful to the following for permission to reproduce much of the material in this book. Arthur Koestler for his reply to my review of *The Act of Creation*; the Editor of the *New Statesman* for the review itself and the article 'Darwin's Illness'; Miss Ruth D'Arcy Thompson and the Oxford University Press for 'D'Arcy Thompson and *Growth and Form*; the Editor of *Encounter* for 'Herbert Spencer and the Law of General Evolution' and 'Two Conceptions of Science'; the Editor of *Mind* for the review of *The Phenomenon of Man* and the Editor of *Nature* for 'A Biological Retrospect'.

B

D'Arcy Thompson and Growth and Form

D'Arcy Wentworth Thompson was an aristocrat of learning whose intellectual endowments are not likely ever again to be combined within one man. He was a classicist of sufficient distinction to have become President of the Classical Associations of England and Wales and of Scotland; a mathematician good enough to have had an entirely mathematical paper accepted for publication by the Royal Society; and a naturalist who held important chairs for sixty-four years, that is, for all but the length of time into which we must nowadays squeeze the whole of our lives from birth until professional retirement. He was a famous conversationalist and lecturer (the two are often thought to go together, but seldom do), and the author of a work which, considered as literature, is the equal of anything of Pater's or Logan Pearsall Smith's in its complete mastery of the *bel canto* style. Add to all this that he was over six feet tall, with the build and carriage of a Viking and with the pride of bearing that comes from good looks known to be possessed.

D'Arcy Thompson (he was always called that, or D'Arcy) had not merely the makings but the actual accomplishments of three scholars. All three were eminent, even if, judged by the standards which he himself would have applied to them, none could strictly be called great. If the three scholars had merely been added together in D'Arcy Thompson, each working independently of the others, then I think we should find it hard to repudiate the idea that he was an amateur, though a patrician among amateurs; we should say, perhaps, that great as were his accomplishments, he lacked that deep sense of engagement that marks the professional scholar of the present day. But they were not merely added together; they were integrally – Clifford Dobell said chemically – combined. I am trying to say that he was not one of those who have made two or more separate and somewhat incongruous reputations, like a composer-chemist or politician-novelist, or like the one man who has both ridden in the Grand National and become an F.R.S.; but that he was a man who comprehended many things with an undivided mind. In the range and quality of his learning, the uses to which he put it, and the style in which he made it known I see not an amateur, but, in the proper sense of that term, a natural philosopher – though one dare not call him so without a hurried qualification, for fear he might be thought to have practised what the Germans call *Naturphilosophie*.

Let me now try to describe the environment in which D'Arcy the scientist lived and worked. When D'Arcy flourished, British zoology, after fifty years, was still almost wholly occupied with problems of phylogeny and comparative anatomy, that is, with the apportioning out of evolutionary priorities and the unravelling of relationships of descent. Comparative anatomy has many brilliant discoveries to its credit; for example, the discovery that the small bones of the middle ear – those which transmit vibrations from the ear drum to the organ of hearing – are cognate with bones which in the remote ancestors of mammals had formed part of the hinges and articulations of the lower jaw; that the limbs of terrestrial vertebrates had evolved along a just discernible pathway from the paired fins of fish; that the muscles which move the eyeballs derive in evolution from the anterior elements of a segmental musculature which at one time occupied the body from end to end. But although later work refined upon them or corrected them here or there, all these discoveries had been made in the nineteenth century. When D'Arcy took his chair, the great theorems of comparative anatomy had already been propounded, and nearly all the great dynasties in the evolutionary history of animals were already known. By 1917, the date of the first edition of D'Arcy's essay *On Growth and Form*, British zoology was far gone in that decline from which a small group of 'comparative physiologists' was to rescue it in the middle 'twenties – in due course (it has been rudely said) to impose upon it a hegemony of their own. The work of phylogeny and comparative anatomy is not yet all done. We are still uncertain about the affinities of

whales, though we may be quite sure that purely anatomical research will not reveal them; the comparative anatomy of the lym-
phatic system has hardly been attempted; and there is doubtless
much to be done among the parish registers of evolution, in the
attempt to trace lines of descent within families of animals, or
even within genera. But comparative anatomy is no longer
thought of as the central discipline of zoology; D'Arcy Thompson
was the first man in this country to challenge its pretensions and
to repudiate the idea that zoological learning consisted of so many
glosses on an evolutionary text.

In D'Arcy Thompson's earliest writings there is little to suggest
that he would one day slough off the coils of evolutionary anatomism, though one of his papers – 'On the Nature and Action of Certain Ligaments,' 1884 – is evidence that he was interested in bones
for how they worked rather than for what they might have to say
about their owners' evolutionary credentials. Later, writing again
of ligaments, he said:

> The 'skeleton', as we see it in a Museum, is a poor and even a
> misleading picture of mechanical efficiency. From the engineer's
> point of view, it is a diagram showing all the compression-
> lines, but by no means all of the tension-lines of the construc-
> tion; . . . it falls to pieces unless we clamp it together, as best we
> can, in a more or less clumsy and immobilized way. In pre-
> paring or 'macerating' a skeleton, the naturalist nowadays carries
> on the process till nothing is left but the whitened bones. But the
> old anatomists . . . were wont to macerate by easy stages; and in
> many of their most instructive preparations, the ligaments were
> intentionally left in connection with the bones, and as part of
> the 'skeleton'.

Whitened immobile bones, rearticulated with bits of wire, were
indeed symbolic of the evolutionary anatomism which had 'all
but filled men's minds during the last couple of generations'.

The treatment of bones and other bodily structures as so many
archives of evolution angered D'Arcy for two chief reasons; first,
because his contemporaries and immediate predecessors (it is

difficult to speak of the contemporaries of a man who lived so
long, but here and elsewhere I centre his life about the year of the
publication of *Growth and Form*) had no real curiosity beyond the
evolutionary pedigree of an organism or an organ: any inquiry
into the action of contemporary physical causes seemed to them to
belong to some other science; and secondly because the comparative anatomists, in spite of their devotion to the study of its consequences, were no more than idly curious about the *mechanism*
of evolution; they accepted the contemporary and far from adequate form of Darwinism in much the way that nicely brought
up people often accept their religion, that is, in a manner that con-
trives to be both tenacious and perfunctory. D'Arcy's own opinion
was that we should look to the action of contemporary and immediately impingent physical causes for the explanation of an
animal's growth and form. What causes the spicules of sponges to
take their characteristic shapes? D'Arcy sought the answer in the
phenomena of crystallization and of adsorption and diffusion, instead of being content with the explanation for which he makes
Minchin spokesman, viz. that 'The forms of the spicules are the result of adaptation to the requirements of the sponge as a whole,
produced by the action of natural selection upon variation in every
direction,' and that their regular form is 'a phylogenetic adaptation,
which has become fixed and handed on by heredity, appearing in
the ontogeny as a prophetic adaptation'. For sponges and spicules
one could substitute other organisms and other organs: the formula would accommodate all comers. Then again, D'Arcy tried
(very imperfectly to be sure) to envisage how the physical forces
acting upon them might have shaped the shells of the Foramini-
fera, instead of being content to see in their diverse patterns no
more than the branches of a hypothetical family tree. In embryology, evolutionary anatomism seemed particularly inexcusable,
for real embryos, unlike hypothetical ancestors, are tangible present
objects, and amenable therefore to 'causal' investigation in the sense
in which the physicist employs that term. Yet Hertwig declared
for *post hoc, propter hoc*, holding that the chronological anatomy of embryos provided causes sufficient to explain develop-

ment, and Balfour valued embryology mainly for its testimony of descent. Of course, it was not D'Arcy himself, but long before him Roux and His who founded modern analytic embryology by trying to introduce a little dynamics into evolutionary dynastics; but for so long did the spirit of anatomism prevail that even when I was a student at Oxford, the causal analysis of development was separated from descriptive embryology and treated as a thing apart. No wonder D'Arcy was an anti-Darwinian! Believing as he did that present phenomena should be explained by present causes, he saw the appeal to deep historical antecedents as an evasion of responsibility – all the more culpable for being made with the authority of what was, at the time, a most imperfect evolutionary theory.

Here I believe that D'Arcy was as much in error as those whose doctrines he endeavoured to correct. I must make my point at length, because it is the substance of the charge that D'Arcy was somehow 'unbiological' in his way of thinking, and it explains why, although he was surely right to be annoyed with his more austerely anatomical colleagues, they in their turn were not wholly to blame for feeling annoyed with him.

Consider the argument set out in Chapter XVI, 'On Form and Mechanical Efficiency', in the essay *On Growth and Form*. Here D'Arcy tells us, amongst other things, of the fitness to their several purposes of bone and bones. The shafts of the humerus and femur are hollow cylinders of a dense compact bone which is thicker in the middle than at either end, for it is sound engineering to have the thickness vary with the bending moments. At each end the shafts widen, and the compact bone thins out, enclosing within it a thick layer of bone of a more open texture – cancellar bone, made of interesting laminae or trabeculae of bony matter. In a section cut lengthwise through the head of, for example, the human femur, one can see that the bony trabeculae are not arranged haphazardly, but that they form two systems of curves, intersecting roughly at right angles, which are (to a fair approximation) a structural embodiment of the system of stresses to which the bone is exposed in normal life. (It is an old story this, which goes back to Julius Wolff and Herman Meyer.) The entire arrangement clearly represents an adaptation: what kind of adaptation could it be?

The trabeculae, D'Arcy reminds us, are not permanent structures: they are constantly being broken down and formed anew, and if by mischance a bone should be broken and should reunite in some abnormal fashion, the trabeculae will shape themselves into a new pattern governed by the new and altered system of stresses and strains. This, then, should give us almost direct evidence of what must happen in ordinary development: the trabeculae begin by being 'laid down fortuitously in any direction within the substance of the bone' but end in that functionally apt pattern which seems so clearly to represent the engravery of actual use. 'Here, for once, it is safe to say that "heredity" need not and cannot be invoked to account for the configuration and arrangement of the trabeculae: for we can see them, at any time of life, in the making, under the direct action and control of the forces to which the system is exposed.' D'Arcy could have drawn the same conclusion from a study of the forms of joints, for when a long bone is broken and the broken ends fail to reunite, they sometimes become hinged to each other in what is anatomically an almost perfect joint. What better evidence could there be that joints too, with all the niceties of their patterns of articulation, are shaped by use?

Yet they are not so. For all their fitness to mechanical purposes, the patterns of bone and bones are not, in the first instance, moulded by the demands of use; the evidence of remodelling and regeneration shows that they *could* be so, and that under special circumstances they are so; but bones will develop in an anatomically almost perfect fashion even when deprived of innervation or transplanted into positions where they can neither move nor be moved. To explain the shapes of bones we must look elsewhere than to the mechanical forces that act in an individual's own lifetime. We need not doubt that D'Arcy's forces acted once upon a time in directing the pathway of evolution, but if that is so, then the problem is just what D'Arcy supposed it not to be: a matter of history. The whole point is that the forces which did at one time shape limbs, or set the limits within which the shapes of limbs must fall, were translated into those developmental *instructions* about limb-making that now form part of our genetic heritage; the problem of the development of limbs is, first, to break the chemical code which embodies the instructions, and second to find out how the instructions take effect. Here, too, lies the sense of that venerable old antithesis between 'preformation' and 'epigenesis': the instructions are ready-made, but their fulfilment is epigenetic: heredity proposes and development disposes. No one denies the essential truth of what D'Arcy Thompson had to say about the influence of physical and mechanical forces; he simply mistook their context. Perhaps he was aware of having done so when, on a later page, he admits that 'a principle of heredity' may have much to do with the matter. But he never quite realized that he and the comparative anatomists were giving not rival answers to the same question but different answers to two different questions. D'Arcy's answer is to the question: What physical agencies formed the basis of natural selection, and so caused one particular set of instructions about limb-formation to prevail? So construed, all that he has to say is relevant. After all, the elementary forces whose action he contemplated were no different in any yesterday from what they are at present: 'a snow crystal is the same today as when the first snows fell'.

Because he spoke out impatiently against contemporary orthodoxy, D'Arcy Thompson is often thought of as a great innovator; but the angle subtended between an innovator and his contemporaries gets smaller in the more distant view, and at a distance of forty years it no longer seems to us that he stood so far to one side of his anatomical colleagues. Anyone who reads *Growth and Form* attentively will soon discern that D'Arcy was an anatomist himself; the life that appears in his pages is usually still life, and it was product rather than performance that usually engaged his attention. I am well aware that his conception of form was essentially dynamical; but though he did indeed declare that any given shape was to be thought of in terms of some orderly array of inequalities of growth-rates, yet in his 'Method of Transformations' (which I shall mention later) he did in fact compare the final products of two processes of transformation and not the processes themselves.

If D'Arcy was an anatomist, he was the first completely modern anatomist, in that his conception of structure was of molecular as well as of merely visible dimensions, his thought travelling without impediment across the dozen orders of magnitude that separate the two. The advances that have occurred in modern biophysics and structural biochemistry are comprehended within D'Arcy's way of thinking. We all now understand, though the idea was revolutionary in its day, that molecules themselves have shapes as well as sizes: some are long and thin, others broad and flat, some straight, some branched. We also know that crystalline structure is enjoyed by the huge molecules of proteins, nucleic acids, and poly saccharides as well as by the 'crystalloids' of an older terminology, and that the structure of the cell surface and of some of the 'organelles' enclosed by it must be interpreted in molecular terms. D'Arcy would have delighted in the modern X-ray crystallography of 'biological' compounds and in the penetrating insight of the higher powers of the electron microscope; particularly would he have rejoiced in the modern solution of the structure of desoxyribonucleic acid – a brilliant feat of chemical anatomy that provides us for the first time with some physical conception of the 'gene'. Biology, and the chemistry and physics that go with it, have grown more rather than less anatomical in recent years, and the anatomy is indeed Thompsonian, one which recognizes no frontier between biological and chemical form.

This is the right moment to explain D'Arcy's own 'philosophy' of living organisms. He believed (*a*) that the laws of the physical sciences apply to living organisms, and (*b*) that living organisms do nothing to contravene those laws. These propositions are sometimes taken to import that biology is, or soon will be, nothing more than a kind of super physics-and-chemistry. In reality they do nothing of the kind. Biology deals with notions that are contextually peculiar to itself – with heredity, development, and sexuality; with reflex action, memory, and learning; with resistance to disease and disease itself. These things are no more part of physics and chemistry than is the Bank Rate or the British Constitution. We are mistaking the direction of the flow of thought when we speak of 'analysing' or 'reducing' a biological phenomenon to physics and chemistry. What we endeavour to do is the very opposite: to assemble, integrate, or piece together our conception of the phenomenon from our particular knowledge of its constituent parts. It was D'Arcy's belief, as it is also the belief of almost every reputable modern biologist, that this act of integration is in fact possible. He stopped short of supposing that the act of integration would eventually irrupt upon matters of the spirit: 'Of how it is that the soul informs the body, physical science teaches me nothing: . . . nor do I ask of physics how goodness shines in one man's face, and evil betrays itself in another.' But D'Arcy makes no other mention of these matters, and nor shall we.

The essay *On Growth and Form* (so he described it, though it is nearly 800 pages long) was D'Arcy's magnificent attempt to put his philosophic principles to work. The attempt was successful in so far as it depended upon his geometrical insight, and courageous (though inevitably often faulty) whenever the physicist got the upper hand. The biologist in him, strangely enough, was the weakest member of the team. D'Arcy's treatment of Form, then, is generally illuminating, particularly when he tells us how pervasive are certain elementary forms – the unduloid and catenoid, as well as their simpler relatives the sphere and cylinder; the logarithmic spirals of the horns and shells that grow by accretion while remaining unchanged in shape, the geodetic lines of the thickenings of plant cell walls, the geometric packing adopted by cellular aggregates. When it came to the physicist's turn, and the attempt to explain the shapes of cells or of spicules, or the mechanisms of amoeboid movement or phagocytosis, then indeed his simple armament of surface tension, viscosity, diffusion, and adsorption was not powerful enough by far. We cannot reproach D'Arcy for having failed to solve problems which, most of them, defy us yet; but it is a fair comment that D'Arcy himself made no move to solve them, whether by doing experiments or by suggesting experiments that might have been done by others.

D'Arcy Thompson was sometimes accused of being too much the geometer in his way of thinking, because of his determination to see simple regularity where a literal-minded person would say it did not exist: the spheres he saw were not quite spherical, the polygons not quite regular, the transformations not quite orthogonal, the bony trabeculae an inaccurate representation of stress and strain. It is an old and wearisome story, and in finding this new context for it I believe that D'Arcy's critics were completely wrong. Surely we must always begin by seeking regularities. There is *something* in the fact that the atomic weights of the elements are so very nearly whole numbers; that there is a certain periodicity in the properties of the elements when they are written down in the order of their ascending weights; that hereditary factors tend to be inherited not singly but in groups equal in number to the number of pairs of chromosomes. Unless we see these regularities and strive to account for them, we shall not equip ourselves to understand or even perhaps to recognize the existence of isotopes, the significance of atomic number, or the phenomenon in genetics of crossing over. We act either as D'Arcy does, simplifying and generalizing until the facts confute us, or, like those in whom Bacon saw a chief cause of the retardation of learning, we must betake ourselves to 'complaints upon the subtleties of nature, the secret recesses of truth, the obscurity of things, and the infirmity of man's discerning power.'

D'Arcy must be acquitted of the charge of striving without just reason for simplicity, but the charge that the biologist in him was strangely unperceptive cannot so easily be dismissed. Here is an example taken from *Growth and Form* in which even a layman should be able to see a certain perversity of reasoning. It comes from Chapter II, 'The Rate of Growth.'

To us nowadays, as fifty years ago to the Bostonian anatomist Charles Minot, it seems obvious that the *norm* of biological growth – the standard to which all actual instances of growth must be re-

ferred – is growth by compound interest, sometimes called logarithmic or exponential growth. A house grows by accretion:
the bricks which represent the unit increments of its growth stay
put and do not grow themselves. But the central characteristic
of biological growth is that that which is formed by growth is
itself capable of growing: the interest earned by growth becomes capital and thereupon earns interest on its own behalf.
We must therefore plot the growth of organisms not against an
ordinary arithmetic scale, equal subdivisions of which represent equal additions of size, but against a logarithmic scale, of
which equal subdivisions represent equal *multiples* of size. The
simplest case is when the interest earns interest at the same
rate as the capital to which it is added; in such a case, the loga-
rithm of size will give a straight line when plotted against
age. D'Arcy missed the point of Minot's appeal that this
was the form in which curves of biological growth should be represented, just as, later, he was to miss the point of J. S. Huxley's
analogous treatment of the phenomenon of differential growth,
in which the multiplication-rate of one part of the body is measured against the rate of multiplication of another. D'Arcy pre-
ferred to deal with the arithmetical or 'simple interest' type of
growth curve, with its first derivative, growth-rate, and its
second derivative, acceleration of growth. This treatment not
only hid from him much that was important, but led him to attach
importance to things of no great weight – for example, the time of
life at which the arithmetic growth-rate is at a maximum, seen as
the point of inflexion of the integral curve of growth.

   The best and most famous chapter of *Growth and Form* is that
which embodies D'Arcy Thompson's own most completely
original contribution to biology: the 'Method of Transformations'
in the chapter 'On the Comparison of Related Forms'. Consider
the shapes of two organisms of related genera. Being closely related, the two shapes will be 'homeomorphic' – that is, roughly
speaking, they will be qualitatively the same, so that the one could
be changed into the other, as two different faces could be, by some
process of plastic remodelling. But although two such shapes may

be qualitatively similar, they will in fact differ in a multitude of particular little ways. The anatomist's method of comparing shapes is to do so piecemeal, feature by feature, with perhaps an occasional measurement of proportion. D'Arcy saw that all these particular little differences of disposition, angle, length, and ratio might be simply the topical expressions of some one comprehensive and pervasive change of shape. He grasped the transformation *as a whole*. So it might be, for example, with the change of shape produced by distorting a sheet of rubber on which has been drawn a house, or face, or any other figure whatsoever: the shapes of the drawings change in every single particular, but the transformation as a whole might be defined by some quite simple formula describing the way the rubber had been stretched. So also with the transformation that may be produced by tilting a lantern screen away from its normal perpendicular position; whatever is shown on the screen will be transformed in every detail, but in a manner which can be summarily defined by formulae expressing the direction and the degree of tilt. In all such cases the best way of grasping the transformation is to take, as its subject, some simple and regular figure; for mathematical purposes one takes the ordinary 'Cartesian grid' of graph-paper, an orthogonal system of equally spaced straight lines. This was the way in which D'Arcy expressed the relationship between *Diodon* and *Orthagoriscus* or between *Argyropelecus* and *Sternoptyx* (Fig. 1), to choose two among the best of a number of good examples. The grids superimposed upon these figures give one instantly the sense and trend of the transformations; we may think, if we like, that *Orthagoriscus* is a *Diodon* living in some quite remarkably non-Euclidian principality of the ocean, or that *Diodon* is an *Orthagoriscus* of ordinary Cartesian seas. The lesson to be learnt is that we do not have to seek a hundred different explanations of the hundred particular differences between the one form and the other; *one* system of 'morphogenetic forces' may perhaps account for all the differences between the two.

It is hardly necessary to say that the whole treatment is an oversimplification, and that the transformations figured by D'Arcy could not conceivably have happened in real life. In real life, one adult does not change into another adult, but two related embryos turn into two related adults. However, D'Arcy's somewhat elliptical treatment does not make his one important lesson any the less

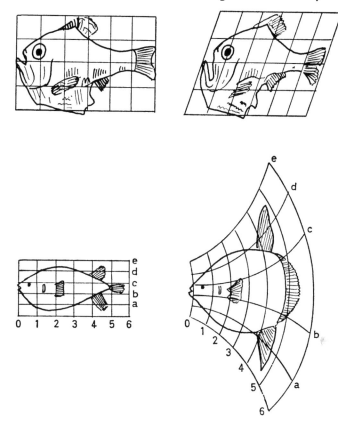

Fig. 1.—*Upper left,* Argyropelecus; *upper right,* Sternoptyx; *lower left,* Diodon; *lower right,* Orthagoriscus.

clear. The reason why D'Arcy's method has been so little used in practice (only I and one or two others have tried to develop it at all) is because it is analytically unwieldy. The methods later developed by J. S. Huxley, though far less comprehensive and ambitious, were much more usable and informative, and they were

c

widely taken up from the moment they were first described. D'Arcy Thompson gave rather perfunctory praise to these developments, believing Huxley's methods of analysis to be implicit in his own. It is in the 'new edition' of *On Growth and Form*, of 1942, that D'Arcy hints at this opinion; I have been writing all the time of the edition that came out at the corresponding period of the First World War. D'Arcy's reputation as a scientist rests almost wholly upon his 'old' edition, and some of D'Arcy's colleagues thought him unwise in his attempt to bring it up to date.

There is one more thing to be said of *Growth and Form*, and perhaps it is the most important thing of all; it relates to an accomplishment in which D'Arcy himself took the utmost pride, and in which he knew in his heart he had no equal. I think that *Growth and Form* is beyond comparison the finest work of literature in all the annals of science that have been recorded in the English tongue. There is a combination here of elegance of style with perfect, absolutely unfailing clarity, that has never to my knowledge been surpassed. To be sure, much of D'Arcy's writing sounds old-fashioned – his prose has a longer stride than we can keep up with nowadays – and the scholarly allusions and digressions and the little graces of writing may be somewhat overdone; but this and all the other decorative matter is simply the *fioritura* of the perfectly accomplished singer. *Growth and Form* will remain for ever worth reading as a text in the exacting discipline of putting conceptions accurately into words.

The influence of *Growth and Form* in this country and in America has been very great, but it has been intangible and indirect. It is to be seen in anyone who, having read it, tries to write a little more clearly and with at least an attempt at grace; or who realizes, perhaps for the first time as he turns its pages, that science cannot be divided into what is up to date and what is merely of antiquarian interest, but is to be regarded as the product of a growth of thought. Most clearly of all is to to be seen in the complete matter-of-factness with which we now accept certain beliefs that D'Arcy, as a natural historian, had to fight for: not merely that the physical sciences and mathematics offer us the only pathway that leads to an understanding of animate nature, but also that the true beauty of nature will be revealed only when that understanding has been achieved. To us nowadays it seems obvious that the picture we form in our minds of Nature will be the more beautiful for being brightly lit. To many of D'Arcy's contemporaries it must have seemed strange and even perverse that he should have combined a physico-mathematical analysis of Nature with, at all times, a most intense consciousness of its wonder and beauty; for at that time there still persisted the superstition that what is beautiful and moving in Nature is its mystery and its unrevealed designs. D'Arcy did away for all time with this Gothick nonsense: a clear bright light shines about the pages of *Growth and Form*, a most resolute determination to unmake mysteries.

It is by such diffused and widely pervasive effects as these that we must measure the influence of *Growth and Form* upon biological science. Of direct influence, that can be traced in pedigrees of teaching or research, there is little. In a generation's time there will be no one alive who heard D'Arcy's lectures, and no one to declare from personal knowledge that he knew the animal kingdom inside out. D'Arcy had only one pupil of more than ordinary distinction, and *he* made his name in descriptive zoology of a most un-Thompsonian kind. D'Arcy founded no school, as Sherrington did, so that no lineage of research can be traced back directly to sources in his mind. But then, he did no research in the modern sense; he was, as I said to begin with, a natural philosopher, one who by reflection rather than by intervention or experiment arrived at a certain imperfect but nevertheless whole conception of that science in which God has been slowest to reveal Himself a geometer. It was a conception expressed for the most part in a modern scientific idiom, and with a beauty and clarity of writing that may never be surpassed.

*Herbert Spencer and the Law of
General Evolution*

I hesitated some little while before accepting Oxford's invitation to deliver this year's Herbert Spencer Lecture. My conscience told me that I could not honourably accept unless I were prepared to steep myself in the work and thought of Spencer himself, an enterprise not to be lightly undertaken. For Spencer was a System Philosopher, one who endeavoured (in Whitehead's words) 'to frame a coherent, logical, necessary system of general ideas in terms of which every element of our experience can be interpreted.'[1] And his System was set forth in twelve volumes thicker and squarer than Gibbon's, each bound in a cloth which has acquired with age a reptilian colour and texture, so putting one in mind of some great extinct monster of philosophic learning.

The prospect did not beckon me on; nor was I cheered up by finding myself the first man ever to have read the two volumes of the *Principles of Biology* acquired by the Royal Society's library more than half a century beforehand. But when I began to read I knew that the challenge was not to be resisted. I began to understand why in his lifetime Herbert Spencer's work had sold in tens of thousands. The *Study of Sociology* at 10s. 6d. had sold more than 20,000 copies by 1900, and the cheap edition of his tract on *Education* nearly 50,000 – and these, it should be remembered, lay outside his formal *System of Synthetic Philosophy*. The System comprised the Principles of Sociology, and of Psychology, Biology, and Ethics, the whole knit together by a volume of *First Principles*, 'primordial truths' arrived at by deduction from 'the elementary datum of consciousness'. What a tremendous undertaking! And what a formidable man Spencer was! His energy was apparently equal to any exertion; his thought had a steady pounding forward motion along the lines he had laid down for it in the

[1] A. N. Whitehead, *Process and Reality* (Cambridge, 1929).

famous manifesto or prospectus that preceded the first edition of the *First Principles*.[1]

I think Spencer was the greatest of those who have attempted to found a metaphysical system on naturalistic principles. It is out of date, of course, this style of thought; it is philosophy for an age of steam; and until a few years ago we should have been tempted to describe it as equally out of date in content. Evidently it is not. Spencer's greatest contribution to philosophy was his Theory of General Evolution, which I shall expound in a moment, and in recent years his ideas have come back to life, or been propped upright again, in the work of men as far apart as Julian Huxley and Father Teilhard de Chardin,[2] to say nothing of the revival of evolutionary sociology and social anthropology. I intend it to be a compliment to both parties when I say that Huxley's thought about evolution is in the same general style as Spencer's. Teilhard, on the contrary, was in no serious sense a thinker. He had about him that innocence which makes it easy to understand why the forger of the Piltdown skull should have chosen Teilhard to be the discoverer of its canine tooth.[3]

Now what I propose to do is to show that the principle of general evolution is not an important principle, and that abstractions arrived at in the way this one was arrived at have the property of sounding as if they were tremendously 'significant' without actually being so. I then turn to a much more important problem arising directly out of Spencer's evolutionism, namely the difficulty he felt obscurely (and others have since felt very clearly) of reconciling the Law of General Evolution with another great natural law – one that pronounces for a general decay of order and a great levelling of energy, and declares that the direction of the flow of natural events is always twards what Willard Gibbs called

[1] *Le style est l'homme même.* I had quite forgotten, when I wrote this, that Spencer began his professional life as a railway engineer.

[2] See *The Phenomenon of Man* (London, 1959).

[3] On August 30th, 1913. The whole story is to be found in *The Earliest Englishman* by A. S. Woodward (London, 1948); see also Charles Dawson and A. S. Woodward, *Quart. J. Geol. Soc.* (London, 70 pp. 82–99, 1914.

mixedupness. I shall try to identify the various misunderstandings which have led to the belief that living organisms circumvent or actually break this Second Law of Thermodynamics, and shall then argue that the equation of biological order or organization with thermodynamic order, and so in turn with information content and the idea of improbability, is one that cannot be sustained.

To show that I have no grudge against evolution as such, and that I accept the Laws of Thermodynamics in the spirit in which Carlyle's lady correspondent accepted the Universe ('by Gad she'd better!'), I emphasize that I shall say nothing about the principle of General Evolution that I would not be prepared to say about any other attempt to pass off a mere inductive *collage* as a work of philosophic art.

Consider, for example, a great new universal Principle of Complementarity, (not Bohr's) according to which there is an essential inner similarity in the relationships that hold between antigen and antibody, male and female, electropositive and electronegative, thesis and antithesis, and so on. These pairs have indeed a certain 'matching oppositeness' in common, but that is *all* they have in common. The similarity between them is not the taxonomic key to some other, deeper affinity, and our recognizing its existence marks the end, not the inauguration of a train of thought. The several manifestations of complementarity are so completely different in origin, nature, and import that the properties of the one pair need teach us nothing about the properties of any other. What we do learn is to recognize the relationship when it turns up in a new and unfamiliar context. The idea of complementarity has, for example, never been far from the thoughts of those who have tried to find out how two chromosomes come to be formed where there was only one before; and for biologists the quintessential example of complementarity is indeed the relationship between the twin strands of the molecule of deoxyribonucleic acid, D N A.

Spencer was the first great evolutionist, and he gave the word *evolution* its modern connotation in English. His first account of the matter is in his 'Development Hypothesis,'[1] a (for Spencer) relaxed and fairly chatty argument that appeared in *The Leader* between 1851 and 1854; that is, seven years before the publication of the *Origin of Species*, and when Spencer himself was in his early thirties. In it Spencer asks why people find it so very difficult to suppose 'that by any series of changes a protozoon should ever become a mammal' while an equally wonderful process of evolution, the development of an adult organism from a mere egg, stares them in the face. We can tell from the tone of his article that evolution was already an idea widely discussed by people of philosophic tastes.

As his thought developed Spencer came to think of genetic evolution, evolution in Darwin's sense, as no more than one manifestation of a far grander and more pervasive process; and out of this conviction his System grew. Today we realize that philosophers devise Systems because it gives them a nice warm comfortable feeling inside; it is something done primarily for their benefit, not for ours. Spencer would not have taken kindly to such an interpretation. Nor did he believe that his concept of general evolution grew inductively out of the contemplation of its several instances. On the contrary: Spencer, like Whitehead after him (the last of the great system philosophers) undertook 'the deduction of scientific concepts from the simplest elements of our perceptual knowledge'.[2] *First Principles* was an attempt to do just this: to show that the concept of general evolution followed 'inevitably' from laws of the indestructibility of matter and of the conservation of energy. Spencer's argument is unimportant and unconvincing, its sole purpose being to justify his expectation of finding evolution at work everywhere. The Universe evolved, and the solar system and earth within it. Animals and plants evolve generation by generation, and within any one generation the development of each individual is itself an evolution. Society is an organism and society evolves. Moreover, 'the law of evolution holds of the inner world as it does of the outer world'. Mind evolves; and language and musical expression, the plastic arts and the arts of

[1] Reprinted in *Essays: Scientific, Political and Speculative* (London, 1868).

[2] *An Enquiry concerning the Principles of Natural Knowledge* (Cambridge, 1919).

narrative and dancing, all display one characteristic or another of evolutionary change. Evolution is 'a universal process of things'. And when we contemplate it as a whole, in its 'astronomic, geologic, biologic, psychologic, sociologic, etc.', manifestations,

> ... we see at once that there are not several kinds of Evolution having certain traits in common, but one Evolution going on everywhere after the same manner. . . . So understood, Evolution becomes not one in principle only but one in fact.[1]

These larger ideas, I should explain, grew upon Spencer during the latter part of his life. I am quoting from the last edition of *First Principles*; they are not to be found in the first edition of 1862.

What then was this universal law of the transformation of matter and energy? He picked his way towards a definition or description that satisfied him, but even after forty years he was wanting to polish and qualify it still. What he has to say about definition itself, the process of defining, is an example of his splendid good sense and of his powerful, hideous prose – the writing of a man who, lacking and perhaps contemptuous of the stylistic graces, is absolutely determined to be understood:

> A preliminary conception, indefinite but comprehensive, is needful as an introduction to a definite conception. A complex idea is not communicable directly, by giving one after another its component parts in their finished forms; since if no outline pre-exists in the mind of the recipient these component parts will not be rightly combined. Much labour has to be gone through which would have been saved had a general notion, however cloudy, been conveyed before the distinct and detailed delineation was commenced.

The point is commonplace nowadays, but many scientists still persist in the belief that no rational discourse is possible unless one 'defines one's terms'.

In the outcome, Spencer's definition, as it is to be found in the final revise of *First Principles*, ran thus:

[1] *First Principles* (6th Edition, revised 1900), §188.

Evolution is an integration of matter and concomitant dissipation of motion; during which the matter passes from an indefinite, incoherent homogeneity to a definite, coherent heterogeneity; and during which the retained motion undergoes a parallel transformation.

At once he goes on to say, and we love him for it:

NOTE.—Only at the last moment, when this sheet is ready for press and all the rest of the volume is standing in type . . . have I perceived that the above formula should be slightly modified . . . by introduction of the word 'relatively' before each of its antithetical clauses.

What his Principle of General Evolution amounts to is this: that the direction of the flow of events in the Universe is from simple to complex, diffuse to integrated, incoherent to coherent, independent to interdependent, undifferentiated to differentiated; from homogeneous and uniform to heterogeneous and multiform; and from an abundance and confusion of motion to a regimentation and loss of motion. These are mostly Spencer's own words. He does not speak of a passage from randomness to orderliness, or from more probable to less probable configurations of matter; but if these antitheses had been put to him, I feel sure he would have accepted them as fair descriptions of the trend or tendency of evolution.

Let us now study the principle of general evolution in its biological contexts to see if it actually works.

When used without further qualification, the word 'evolution' is generally taken to mean evolution in the genetic or Darwinian sense. Provided we confine ourselves to comparisons between grown-up organisms, evolution of this sort answers to Spencer's definition pretty well: there is indeed a passage from simple to complex and towards a differentiation and mutual dependence of parts. But Spencer's formulation completely fails to cope with the very real and important sense in which a frog's egg or embryo is more highly evolved than, say, a grown-up earthworm. Indeed, Spencer's conception of development (which I shall deal with in a moment) might lead us to think embryos *less* highly evolved than the adult forms of their own ancestors. This difficulty disappears if we take the view that evolution in the genetic sense is an evolution *of* developments – or, more exactly, an evolution of the genetic instructions that constitute the programme or specification of development. The genetic instructions that govern the development of a frog are much more elaborate and complicated than those that govern the development of an earthworm, and for that reason a frog's egg may be considered a more highly evolved object than an earthworm of any age.

But this leads to a paradox. Development itself is the golden example of an evolutionary process as Spencer conceived it to be. His thoughts constantly recurred to the evolution of tree from seed and of infant from 'germinal vesicle' (that is, egg). Unfortunately, development cannot be described as an evolution in the one essential sense in which genetic evolution *can* be so described. I have just said that we can get round the difficulty of being obliged to think an embryonic frog less highly evolved than an adult earthworm by thinking of an evolution of earthworm-making instructions into instructions for making frogs. Development is the carrying out of these instructions; it is a film that sticks faithfully to the book – an evolution, then, only in the sense of a translation, spelling out or 'mapping' of one kind of complexity into another kind of complexity. The adult is 'implicit' in the egg in the sense that one day it will be possible, after determining certain parameters, to read off the constitutional properties of the adult animal from a detailed knowledge of the chemical structure of the egg it arose from. Genetical evolution is entirely different; it is not a process of unfolding, and there is no useful sense in which the structure of a mammal can be said to be implicit in the structure of a protozoon.

I cannot make up my mind whether Spencer grasped this point or not. As early as 1852, in *The Development Hypothesis*, he wrote:

The infant is so complex in structure that a cyclopedia is
needed to describe its constituent parts. The germinal vesicle is
so simple that it may be defined in a line.

Forty-five years of reflection must have confirmed him in this
opinion, for the same sentences occur word for word in the first
volume of the revised edition of *The Principles of Biology*. Yet in
that same volume, grappling with the problem of how a peacock's tail comes to acquire its elaborate pattern, he made a rudimentary attempt to estimate what would now be called the amount
of 'information' that must be present in a peafowl's egg to specify
the pattern of one feather of the adult's tail. By erroneous reasoning[1] he came to the conclusion that 480,000 Weismannian 'deter-
minants' would be required to specify the pattern of one feather
alone. No wonder he declared that the 'organizing process trans-
cends conception. It is not enough to say we cannot know it; we
must say that we cannot even conceive it.' (To describe as 'inconceivable' what he himself could not conceive was one of Spencer's
little weaknesses.)

However that may be, I think it must be clear that in describing
both development and phylogenetic transformation as processes
of 'evolution' we may be making a useful statement about one or
about the other; but not, I fear, about both. They are altogether
different phenomena.

Biologists who use English as a scientific language *never* use the
word 'evolution' to describe the processes of growth and development. (In France the usage is different.) They refrain because to do
so would be confusing and misleading.

No such scruples, weigh, alas, with biologists who speak about
'social evolution' or 'psychosocial evolution'. So far as I can make
out from the writings of its various advocates, this superorganic
evolution, as Spencer called it, has many manifestations, of which

---

[1] See *The Principles of Biology* (Revised Edition, 1898), pp. 372–3. The com-
putation is not possible even with the evidence now available to us; but if we
were to attempt it we should certainly not assume that the individual elements of
the pattern behaved as 'independent variables'.

I shall mention only three and discuss only the third. They are: (1) the Spencerian evolution of social organization and of social institutions like governments, joint stock companies, banks, and so on (this was what Spencer himself was mainly interested in). (2) What A. J. Lotka,[1] called the 'exosomatic' (as opposed to ordinary or 'endosomatic') evolution of new sensory organs like spectacles, ear-trumpets, and ultraviolet spectrophotometers, or new motor organs like cutlery and guns. Spencer had some pointed and sensible remarks to make about evolution of this kind in his *Principles of Psychology*. These first two kinds of superorganic evolution are, of course, the consequences of a third: (3) cultural or psychosocial evolution, the secular accumulation of fact and fancy, knowledge and know-how, rules and rites, that is mediated through tradition.

By 'tradition' I mean 'the transfer of information through non-genetic channels from one generation to the next'.[2] I discussed psychosocial evolution at some length in the last of my Reith Lectures,[3] and tried to explain in just what sense psychosocial evolution represents a fundamentally new biological stratagem. But although psychosocial evolution is immensely important, it is also immensely obvious. Spencer did not ration himself austerely where explanations and examples were called for, but of the changes caused by the prodigious secular growth of the arts and sciences he says only 'the proposition is familiar and admitted by all. It is enough simply to point to this great phenomenon as one of the many forms of evolution we are tracing out.'[4] He spent much more time, unfortunately, in trying to demonstrate that the exercise of the mind had a direct hereditary effect on the capabilities of the brain in later generations. Perhaps this is why he believed in the inevitability of progress; for if his interpretation were true, social evolution would be cumulative and virtually irreversible. *We* know better: that we are all born into the Old Stone Age and in principle could stay there.

[1] *Human Biol.* 17: 167 (1945).

[2] *The Uniqueness of the Individual* (London, 1957), p. 141.

[3] *The Future of Man* (London, 1960).

[4] *The Principles of Psychology* (4th Edition, 1899), Vol. 1, §158.

Psychosocial evolution differs from ordinary genetic evolution in three important ways: it is not mediated through genetic agencies; it is reversible, in the sense that what it has gained can in principle be wholly lost, and in one generation; and it is an evolution in the Lamarckian style, in the sense that a father's particular knowledge and skills and understanding can indeed be transmitted to his son, though not (as Spencer supposed) through genetic pathways. Common sense suggests that differences of this magnitude should be acknowledged by a distinction of terminology. The use of the word 'evolution' for psychosocial change is not a natural usage, but an artificial usage adopted by theorists with an axe to grind. If by any chance it *had* been a natural usage, people like myself on occasions like this would have said over and over again how wrongheaded it was, and how wise we should be to abandon it.

All who think about psychosocial 'evolution' agree that its inception marks a second great epoch of biological history. But I wonder: is it a second, or is it perhaps a third? The first must surely have been an evolution at the chemical or molecular level of integration – a process of which we can have no direct knowledge, for in a certain important sense all chemical evolution in living organisms stopped millions of years before even our faintest and most distant records of life began. So far as I know, no new *kind* of chemical compound has come into being over a period of evolution that began long before animals became differentiated from plants. Nor has there been any increase of chemical complexity; no chemically definable substance in any higher organism, for example, is more complex than a bacterial endotoxin. I have no views on the processes of evolution that brought new kinds of chemical compounds into existence, but I should not be surprised to find them very different from the forms of evolution that have been in progress since.

The point I wish to make is that evolution since those primordial days has been an evolution of structure at a higher level of integration than the chemical (using the word 'chemical' in the way it is used by chemists). It was these thoughts that led me to the discussion that now follows on the relationship between biological and thermodynamic order.

The sixteenth chapter – in effect the last – of the first edition of *First Principles* contains an argument on the phenomenon of equilibration from which Spencer ultimately drew

> . . . a warrant for the belief, that Evolution can end only in the establishment of the greatest perfection and the most complete happiness.

This sentence cannot be found in the latest version of *First Principles*. Its place is taken by some sombre and, I must also say, rather confused reflections upon the ultimate state of the Universe: for Spencer now enlarges and develops an argument of which, in the first edition, we see only the embryonic rudiments – an argument tending to the conclusion that unless something unforeseen and unforeseeable turns up, all things must 'beyond doubt' tend towards a universal quiescence, an 'omnipresent death'. What can have been responsible for the much greater weight he gave in his later thought to the phenomena of dissipation and dissolution?

The theory of General Evolution was first hinted at in Spencer's *Development Hypothesis* of 1852. One year before, in the *Transactions* of the Royal Society of Edinburgh,[1] a Professor William Thomson (later Lord Kelvin) called attention to and elaborated upon some 'remarkable conclusions' arrived at by Clausius and Rankine after studying the properties of 'thermodynamic engines', engines that translate heat into mechanical work. The First Law of Thermodynamics (though not then so described) had already brought the reassuring news that heat, as a form of energy, could not be lost, for the total quantity of energy in the universe remained constant, no matter what its transformations. But though not lost to the universe, it now became certain that heat was 'irrevocably lost to man, and therefore "wasted", though not *annihilated*' in thermodynamic transactions; for (Thomson went on to explain) the conversion of heat into mechanical work depends on

[1] Vol. 20 (1851), pp. 261–88.

D

inequalities of temperature within the system, and these inequalities are progressively done away with in a great and universal process of levelling up.

Like the principle of General Evolution, this Second Law of Thermodynamics was in due course taken out of its native environment, here the pithead and the railway workshop, and generalized in much the same way as Spencer generalized the theory of evolution. With the growth of the science of statistical mechanics, it became possible to translate the Second Law into a statement about the history of a system of particles whose behaviour was known in the aggregate only, not individually. This historical statement declares that, in an isolated system, the pattern of the distribution of the elements within it passes from order towards randomness, from separatedness to mixedupness, and, in general, from less probable towards more probable configurations. Our sense of the fitness of things tells us that something is being lost in the process – orderliness, perhaps, or availability of energy or thermodynamic competence – but the historical origins of the concept unhappily still persuade us to speak of a gain of something, of *entropy*, a quantity which, in its native context, is a simple ratio expressing the degree to which thermal energy is no longer available for the execution of mechanical work. '*Die Entropie strebt einem Maximum zu*' was Clausius's own formulation of the Second Law.

The most recent context for the general law of the decay of order and increase of entropy is in the theory of communication. A message encoded in symbols (for example, a Morse signal) owes its specificity, its property of being *this* message and not that message, to the particular configuration or sequence of the symbols, and a random or disorderly configuration of symbols does not make sense. The information capacity of a system of communication obviously depends on the range of different configurations of symbols at the command of the transmitting agent. In a sense, therefore, information capacity is a measure of order or, by a natural extension of the idea, of improbability; information capacity is thus analogous to negative entropy, and may be measured in formally similar terms. This formal similarity has led some people to declare that information *is* negative entropy, but the usage strikes me as perverse. There is much the same formal similarity between the equations for the diffusion of heat and for the diffusion of solutes, but (as I think Hogben somewhere remarked) we nowadays resist the temptation to refer to heat as a caloric fluid. 'Information,' said Professor Norbert Wiener, in a passage more than usually full of negative entropy, 'is information, not matter or energy.' Elsewhere he points out that the concepts of information and of *pattern* are not coextensive: information is a concept normally (if not necessarily) applied to patterns which are spread out or must be read out in a series, in practice a time series.[1] Such is the case with the information in a gramophone or tape record and also, so it now appears, in a chromosome. The order matters.

By the end of the nineteenth century the philosopher could choose between alternative doctrines of world transformation, the one apparently contradicting the other. The principle of General Evolution spoke of a secular increase of order, coherence, regularity, improbability, etc., and Spencer's own derivation made it appear to follow logically from physical first principles; while the Second Law of Thermodynamics, suitably generalized, spoke of a secular decay of order and dissipation of energy.

There can be no doubt that Spencer's thought took on a darker complexion in later years for essentially thermodynamic reasons; but such was the prestige of evolution theory that the Second Law of Thermodynamics was over and over again described as a Law of Evolution, sometimes as *the* law of evolution. A. J. Lotka, the greatest of modern demographers, upheld this interpretation and applied it to biological evolution,[2] but to most

---

[1] *Cybernetics* (New York, 1948), p. 156, and *The Human Use of Human Beings* (London, 1950), p. 21.

[2] *The Elements of Physical Biology* (Baltimore, 1925). Lotka chose to regard evolution as the change undergone by the *total* system 'organisms + environment' conceived as an isolated system (or rather as a closed system with a known input of radiant energy). Conceived thus, the evolving system certainly obeys

people biological evolution and increase of entropy seem mutually contradictory ideas. 'Evolution,' says Julian Huxley,[1] who can be spokesman for all who have thought likewise, 'is an anti-entropic process, running counter to the Second Law of Thermodynamics with its degradation of energy and its tendency to uniformity'; and François Meyer[2] speaks of a principle of Anti-chance at work among living things.

In fact the two concepts are not antithetical. That they are popularly supposed to contradict each other is due in part to a fairly obvious misunderstanding about the physical situations in which the two generalizations hold good and make sense, and in part to a more subtle and correspondingly less obvious confusion of language.

Spencer himself, not unexpectedly, was unable to work the problem out; for this reason the final, revised edition of *First Principles* is much less satisfactory than the first, in which his thoughts were still untroubled by the ideas of universal mixing-up and running-down that grew out of the work of Boltzmann and Willard Gibbs. Spencer always believed that evolution had a natural limit, and came to an end when a certain state of equilibrium was reached, as he believed it always must be – for 'all terrestrial changes are incidents in the course of cosmic equilibration'. His arguments are not very clear because the equilibrium he refers to seems sometimes to mean a state of quiescence and rest, and at other times a steady state in which evolution and its exact opposite, Dissolution, just cancel each other out. But Dissolution eventually supervenes and the universe ends in a ruin of order. Evolution may start up again locally, and perhaps evolution and dissolution may alternate, but the matter must be left open because it is 'beyond the reach of human intelligence'. Spencer's answer seems to have re-

[1] In his *Introduction* to Teilhard de Chardin's *Phenomenon of Man*.

[2] *Problématique de l'évolution* (Paris, 1954).

the second law, and there is much to be said for Lotka's viewpoint. But if we use the word 'evolution' to describe this general transformation, we shall have to invent another word to stand for evolution in its more usual biological sense.

conciled the laws of evolution and of thermodynamics by supposing them to mark not incompatible but successive dynasties in
a history of world order. But this will clearly not do, and we must
seek other more convincing interpretations.

If we confine ourselves merely to considerations of energetics,
there is no problem; or at least, there is no confusion. The Second
Law of Thermodynamics applies to *isolated* systems, systems in
which there is no external trade in matter or energy; the law in no
wise excludes the existence of sub-domains in which entropy may
be decreasing,[1] though necessarily at the cost of a disproportion-
ate increase of entropy elsewhere. The answer that has satisfied
most physicists is that living organisms, thermodynamically *open
systems*, are just such domains. They have been described as 'privileged domains', and this sounds well; though the corollary, that
the rest of the system is underprivileged, sounds rather silly. But if
we contemplate the over-all transformations of matter and energy
within the biological system as a whole, as Lotka did, then of
course nothing happens to contravene the Second Law.

A biologist might still protest that even if living organisms
obey the letter of the Second Law of Thermodynamics, they fail
to observe it in the spirit. I should now therefore like to make a
hasty and perhaps superficial attempt to clarify some of the real
confusions of thought and language that underlie the entire argument. The confusions are genuine, even if my own interpretations
turn out to be inexact.

I spoke a moment ago about the possibility of a grand abstract
equivalence between, on the one hand, entropy, randomness,
probability, and nonsense; and on the other hand between thermodynamic order, 'order' as a biologist might use that word, improbability, and information content. Each of these concepts or
pairs of antitheses has been applied, sometimes critically, more often

---

[1] E.g., in a refrigerator, a heat engine in reverse in which heat flows from a
colder to a warmer environment to increase the temperature of the latter.
Needless to say the flow is far from spontaneous. See A. R. Ubbelohde: *Man
and Energy* (Penguin Books, 1963).

recklessly, to the description of living organisms. Let us therefore consider the sense, if any (1) of regarding biological order as a form of thermodynamic order; (2) of describing organisms in evolution or in everyday life as seats or centres of improbability or Antichance; and (3) of using the concepts and terminology of information theory to describe biological organization. If these correspondences or equivalences should be found faulty then the antithesis between evolution and entropy, using both words in their widest senses, will disappear; for it will turn out that they refer to different properties pertaining to different physical situations.

1. First, then, thermodynamic order and biological order or structural organization. So far as I can make out, all physicists who have considered the matter virtually equate the two, though Eddington[1] had some misgivings. Erwin Schrödinger, in his remarkable little book *What Is Life?* (1944), says that living organisms maintain and add to their state of order by, in effect, feeding on negative entropy; by 'drinking orderliness from the environment' – or, to put it more temperately, by breaking down molecules coming from outside the system to pay the thermodynamic bill for synthesizing molecules within the system. Schrödinger explicitly defines 'order' in the language of energetics, equating it to what was later to be called information capacity.

There must surely be a misunderstanding here.[2] Order or organization as the biologist understands it means complex regularity, with the extra connotation of stability. (By 'regularity' I do not mean symmetry or periodicity; I mean rather 'tidiness' or regimentation, a state of affairs in which each element of the system has its proper place.) Order of this kind is by no means confined to the living world: it is the orderliness of a crystal, of a molecule, or in general of the solid state. But an increase of complex regularity may accompany a *decline* of free energy; for example, in the combination between gaseous hydrogen and oxygen

[1] *New Pathways in Science* (Cambridge, 1935).

[2] See Dr Joseph Needham's searching and thoughtful analysis in his essay on 'Evolution and Thermodynamics' in *Time the Refreshing River* (London, 1943).

to form molecules of water, which are more highly 'organized' in the biological meaning of the word than the parent molecules; or, again, in the phenomena of polymerization and crystallization. In all such cases an increase in the degree of 'organization' accompanies an *increase* of entropy – the opposite of what we should expect if biological and thermodynamic order were essentially the same. Willard Gibbs said entropy was 'mixedupness'; biological order is not, or not merely, unmixedupness.[1]

2. *Probability.* Biologists in certain moods are apt to say that organisms are madly improbable objects or that evolution is a device for generating high degrees of improbability; I have already referred to M. Meyer's views on the workings of a principle of antichance. This is entirely in keeping with the idea that evolution generates negative entropy, because the Second Law of Thermodynamics can be taken to declare that the spontaneous motion of all natural events is from less probable to more probable states.

I am uneasy about this entire train of thought for the following reason. Everyone will concede that in the games of whist or bridge any one particular hand is just as unlikely to turn up as any other. If I pick up and inspect a particular hand and then declare myself utterly amazed that such a hand should have been dealt to me, considering the fantastic odds against it, I shall be told by those who have steeped themselves in mathematical reasoning that its improbability cannot be measured retrospectively, but only against a prior expectation. Name a hand before the deal, I shall be told, and then everyone will be very much taken aback if it turns up.

For much the same reason it seems to me profitless to speak of natural selection's 'generating improbability'. When we speak, as Spencer was the first to do, of the Survival of the Fittest, we are being wise after the event: what is fit or not fit is so described on the basis of a retrospective judgement. It is silly to profess to be thunderstruck by the evolution of organism *A* if we should have been just as thunderstruck by a turn of events that had led to the evolution of *B* or *C* instead. The evolution of *A* was in fact the

[1] Needham, *op. cit.*

most probable net outcome of all the many selective forces that acted on its ancestors. The same goes for artificial selection. If I expose a culture of staphylococci to a certain concentration of penicillin, I *expect* (that is, I attach a high probability to) the evolution of a strain resistant to penicillin. Any other outcome would be improbable and would require a special explanation.

It is not only in their evolution but in their growth and persistence from day to day that organisms have been said to be wildly improbable phenomena. By this is meant, I suppose, that if all the ingredients of the world or the solar system were to be shaken up in a dice box of divine dimensions, the emergence from it of a configuration like an earthworm would be most improbable. How very true! But in the physical circumstances that actually prevail, the growth and maintenance of an organism is the most probable outcome of the events it is taking part in. When I eat a meal, for example, I expect part of it to turn into more of myself. In particular, I shall expect some of it to turn into the chemical substance characteristic of my own blood group, group B. So very highly probable is it that substance B will be formed that a court of law will no longer countenance the possibility of my manufacturing a certain amount of blood group substance A – though substances A and B are almost identical physically and chemically.

Similarly, if I spark a mixture of gaseous hydrogen and oxygen, the most probable consequence is the formation of water, though (as I have suggested) water is a more highly 'organized' substance, as the biologist uses that word, than the elements out of which it was compounded. What *may* be said, I think, is that the conjunction of circumstances of which the formation of water is the most probable outcome is itself very improbable – I mean, the coming together of gaseous oxygen and hydrogen and the application of a spark to the mixture. Certainly organisms, to remain alive, generate improbable, conspicuously non-random situations, and pay a heavy price in energy for doing so; but that is hardly an arresting thought.

3. Finally, 'information'. The ideas and terminology of information

theory would not have caught on as they have done unless they were serving some very useful purpose. It seems to me that they are highly appropriate in their proper context, where we have to do with storing information or sending messages; chromosomes, for example, convey from one generation to the next a message about how development is to proceed, and in speaking of genes and chromosomes the language of information theory is often extremely apt. But I feel we have to be on our guard against treating information content as a measure of biological organization. If it were indeed so then, as Waddington has pointed out,[1] we should be obliged to infer that complexity or degree of organization increased very little in biological development. For development, at all events when it occurs in a nearly closed system, is, from the standpoint of information theory, merely a verbose and repetitious spelling out or biological re-wording of the information encoded in the chromosomes. This is a most unhelpful description of development.

In my opinion the audacious attempt to reveal the formal equivalence of the ideas of biological organization and of thermodynamic order, non-randomness, and information must be judged to have failed. We still seek a theory of Order in its most interesting and important form, that which is represented by the complex functional and structural integration of living organisms. For the present we must be content to say that, in biology, the concepts of entropy and thermodynamic order are appropriate when we are dealing with matters of energetics; of information theory when we are studying how messages are sent and acted upon; and of probability where we are dealing with phenomena that have a random element, as in predicting the outcome of breeding experiments. In our moods of abstract theorization we

[1] 'Architecture and information in cellular differentiation,' in *The Cell and the Organism*, eds. J. A. Ramsay and V. B. Wigglesworth (Cambridge, 1961). For discussions of information theory in biology, see *Information Theory in Biology*, ed. H. Quastler (Illinois, 1953); *The Physical Foundation of Biology*, W. M. Elsasser (New York, 1958).

tend to forget how great and how diverse are the functional commitments of biological macromolecules. They insulate, they fill out; they fetch and carry; they prevent the Organism as a Whole from falling apart or from dissolving in water; they prop up, they protect; they attack and defend; they store energy and catalyze its transfer; they store information and convey messages, and sometimes they themselves *are* messages, The successful prosecution of all these activities depends upon properties more complex, various and particular than can be written down in the language of energetics or information theory.

Where a consortium of brilliant and imaginative theorists has failed, so far, to provide us with the right theoretical equipment for studying biological organization, we need not wonder that Herbert Spencer, working pretty well single-handed, failed too. His System of General Evolution does not really work; the evolution of society and of the solar system are different phenomena, and the one teaches us next to nothing about the other.

Development on the one hand, and the secular transformations of species on the other hand, are indeed both evolutionary processes, but in senses so different that to describe one as an 'evolution' makes it imperative to find some different word to describe the other.

But for all that, I for one can still see Spencer's System as a great adventure, and now that I know my way about those thick, square volumes I do not feel I am taking leave of them for good.

Darwin's Illness

Charles Darwin was a sick man for the last forty of his seventy-three years of life. His diaries tell the story of a man deep in the shadow of chronic illness – gnawed at by gastric and intestinal pains, frightened by palpitations, weak and lethargic, often sick and shivery, a bad sleeper, and always an attentive student of his own woes. His complaints began about a year after his return from the great scientific adventures that occupied the five-year voyage of H.M.S. *Beagle*, and they soon took on a fitfully recurrent pattern. Over the next few years, as he became progressively weaker, Darwin gave up his more energetic pursuits, including the geological field-work he had until then delighted in; and in 1842, when only 33, he and his devoted wife Emma retired to a country house in Kent. Darwin left Down House seldom and England never, relying upon correspondence to keep himself up with scientific affairs, and in later years looking fearfully upon the hubbub that broke out after the publication of the *Origin of Species* in 1859.

Like many chronic invalids Darwin came to adopt a settled routine – now a little walk, now a little rest, now a little reading – and three or four hours' work a day was about all he could find energy for. Yet he looked well enough, and was very far from being disagreeable. Every account makes him out considerate and loving, and his granddaughter, Gwen Raverat, described him as affectionate, spontaneous and gay. Nor had he a feeble constitution. He had been an open-air man, strongly built, and at Cambridge a keen shot and sportsman. His records of the *Beagle* and subsequently of its Master's show him resilient, tough, and full of energy. True, he had not been wholly free of illness. At Valparaiso he had had a fever which, Sir Arthur Keith believed, might well have been typhoid (an important point this, for Keith upheld a psychogenic interpretation of Darwin's later illness); and at Plymouth, waiting fretfully for the *Beagle* to set sail, he complained of palpitations and had dark thoughts of heart disease. But nothing about him gave the slightest premonition of the forty years of invalidism that lay ahead.

What was wrong with Darwin? His own doctors were baffled, and their modern descendants disagree. If Darwin's illness had organic signs they were of a kind his doctors could not then have recognized; they inclined to think him a hypochondriac, and the suspicion that they did so is known to have caused Darwin real distress. Orthodox opinion still has it that Darwin's illness was psychogenic, i.e. arose from causes in his own mind; indeed, it figures in Alvarez's textbook on the neuroses as a type specimen of neurasthenia. But what lay behind his neurotic illness? Alvarez, after fifty years' reflection on the matter,[1] seemed to think poor heredity answer enough, and drew attention to the number of difficult and eccentric Darwin and Wedgwood relatives (Emma was a Wedgwood and so was Charles's mother). For Professor Hubble,[2] Darwin's illness, though beyond question of emotional origin, was a subtle adaptation which protected him from the rigours and buffetings of everyday life, the demands of society and the public obligations of a great figure in the world of learning. 'Darwin by his psychoneurosis secretly and passionately nourished his genius' and so gave himself time to execute his great scientific labours: he 'could have done his work in no other way . . .' There is no refuting Hubble's argument, for there is no argument; the case is presented merely by asseveration ('there can be no doubt', 'it is apparent', 'it is inconceivable', 'it is clear', 'there is overwhelming testimony'). Professor Darlington thinks it possible that the persisting cause of Darwin's illness was the disapproval that grew out of Emma's slow recognition that his doctrines were not such as a Christian might approve of.[3] This, however, is a merely casual suggestion; much more weighty is the full-dress psychoanalytic interpretation of Dr Edward Kempf, one to

[1] *New England med. J.* **261**, p. 1109, 1959.

[2] *Lancet*, **244**, p. 129, 1943; ibid, **265**, p. 1351, 1953.

[3] *Darwin's Place in History* (Oxford, 1959).

which *The Times* felt its readers' attention should be specially drawn.[1]

Kempf believed that Darwin's forty years' disabling illness was a neurotic manifestation of a conflict between his sense of duty towards a rather domineering father and a sexual attachment to his mother, who died when he was eight. His mother, a gentle and latterly an ailing creature, fond of flowers and pets, had propounded a riddle which it was Darwin's life-work to resolve: of how, by looking inside a flower, its name might be discovered. Kempf wrote in 1918 with an arch delicacy that sometimes obscures his meaning, but Good's[2] more recent interpretation leaves us in no doubt. For Good, 'there is a wealth of evidence that unmistakably points' to the idea that Darwin's illness was 'a distorted expression of the aggression, hate, and resentment felt, at an unconscious level, by Darwin towards his tyrannical father'. These deep and terrible feelings found outward expression in Darwin's touching reverence towards his father and his father's memory, and in his describing his father as the kindest and wisest man he ever knew: clear evidence, if evidence were needed, of how deeply his true inner sentiments had been repressed.

> As in the case of Oedipus, Darwin's punishment for the unconscious parricide was a heavy one – almost forty years of severe and crippling neurotic suffering which left him at his very best fit for a maximum of three hours' daily work.

It must be made clear that Darwin's father's tyranny was as unconscious as the hatred it gave rise to. Robert Darwin was a very large (340 lb) and extremely successful Shrewsbury physician who, starting with the £20 given to him by grandfather Erasmus, made a fortune great enough to support all his children in comfort all their lives. He was a rather overbearing man of decided opinions, and we can see an outcrop of his tyrannical inner nature in his reproaching Charles for his idleness and love of sport at Cambridge

---

[1] *The Times*, 31 December 1963. For Kempf, see *Psychoanalyt. Rev.* **5**, p. 151, 1918.

[2] *Lancet*, **I**, p. 106, 1954.

and for his getting into the company of what Charles called 'dissipated low-minded young men'. Robert also thoroughly disapproved of Charles's ambition to join the *Beagle* because he had very much wanted him to go, if not into medicine, then into the church; but later, at least at a conscious level, he withdrew his objections, and it was he who bought Down House for Charles and Emma.

But much else in Darwin's career must have helped to lay the foundations of a lifetime's neurotic illness. Kempf must, I think, have been the first to call attention – obvious though it now seems – to Darwin's intent and continuous preoccupation with matters to do with sex. We need look no further than the titles of his books: *The Origin* itself, of course; *Selection in Relation to Sex; The Effects of Cross- and Self-Fertilization in the Vegetable Kingdom;* and *On the Various Contrivances by which Orchids are Fertilized by Insects.* With so great a load of guilt, need we wonder that at the age of 33 Darwin should have retired from public life to live in quiet seclusion in the country? It was a sacrificial gesture, even a crucifixion: and Kempf calls attention to the inner significance of the fact that it was at the age of 33 that Christ himself was crucified.

What is still more important, I feel, is that psychoanalysis has been able to play a searchlight upon the problem of why Darwin's genius took its distinctive form. Dr Phyllis Greenacre, in her recent Freud Anniversary Lecture,[1] says she suspects that his turning to science was mainly the consequence of a 'reaction to sadomasochistic fantasies concerning his own birth and his mother's death'. But we can be more particular than this. Kempf reveals to us that when Darwin was speculating upon the selection of favourable variations he was thinking, of course, of Mother's Favourites ('Darwin was unable to avoid unconsciously founding his sincerest conclusions on his own most delicate emotional strivings'); and Good explains how in dethroning his heavenly father Darwin found solace for being unable to slay his earthly one.

[1] *The Quest for the Father.* International Universities Press (New York, 1963).

What *was* wrong with Darwin? We may never know for certain, and there is no other testimony to overwhelm us, but Professor Saul Adler F.R.S., of the Hebrew University of Jerusalem, makes a good case[1] for Darwin's having suffered from a chronic and disabling infectious illness called *Chagas' Disease* after Carlos Chagas Sr., the distinguished Brazilian medical scientist who first defined it, and caused by a microorganism whose name, *Trypanosoma cruzi*, honours another distinguished Brazilian scientist, Osvaldo Cruz. Sir Gavin de Beer, in his excellent *Charles Darwin*,[2] thinks Adler's interpretation by far the most likely one. It differs from other theories we have considered in being based upon the use of reasoning, and Adler's case for it runs approximately thus:

On 26 March 1835, when spending the night in a village in the Argentinian province of Mendoza, Darwin was attacked by the huge blood-sucking bug *Triatoma infestans*, the benchuca. The benchuca, the 'great black bug of the Pampas', is the chief vector of *T. cruzi;* even today more than 60 per cent of the inhabitants of Mendoza give evidence of the disease and 'as many as 70 per cent of specimens of *Triatoma infestans* are infected with the trypanosome'. It is very likely, then, that Darwin was infected: South American experts consulted by Adler put his chances of escape no higher than 'negligible'. The symptomatology of Darwin's illness can, it appears, be matched closely by known cases of Chagas' disease in its chronic form. De Beer summarizes the evidence thus:

> The trypanosome invades the muscle of the heart in over 80 per cent of Chagas's disease patients, which makes them very tired; it invades the ganglion cells of Auerbach's nerve plexus in the wall of the intestine, damage to which upsets normal movement and causes great distress; and it invades the auricular-ventricular bundle of the heart which controls the timing of the beats of auricle and ventricle, interference with which may result in heart-block. The lassitude, gastro-intestinal discomfort,

[1] *Nature,* **184**, p. 1102, 1959.

[2] Nelson (London, 1963).

E

and heart trouble from which he suffered an attack in 1873 and died in 1882, all receive a simple and objective explanation if he was massively infected with the trypanosome when he was bitten by the bug on 26 March 1835.

De Beer points out that *T. cruzi* was not identified until twenty-seven years after Darwin's death.

A number of minor and in themselves insubstantial pieces of evidence tell in favour of Adler's interpretation, one of them being that even today inexperienced clinicians may dismiss the chronic form of Chagas' disease as an illness of neurotic origin. I am not aware of any decisive evidence against Adler, and some of the arguments used to discredit his theory establish nothing more than their authors' anxiety to rehabilitate a purely psychogenic interpretation. Disputants so naïve should abstain from public controversy. But clinicians have made it clear to me that the infective theory is by no means a walkover. There is a general and consistent colouring of hypochondria about Darwin's illness; it is a little surprising that we hear nothing about the acute fever and glandular swellings that would surely have followed infection;[1] and so very long-drawn-out a warfare between host and parasite, neither gaining the upper hand for long, is at least unusual. Adler's interpretation has been widely accepted by Brazilian experts, but I suspect that Darwin's having suffered from an illness so closely associated with the names of two great Brazilian scientists is the source of a certain national pride.

The diagnoses of organic illness and of neurosis, are not, of course, incompatible. Human beings cannot be straightforwardly

[1] In this connection, however, my friend Professor P. C. C. Garnham, F.R.S., writes: 'The initial infection in many cases is probably unaccompanied by acute symptoms: there is often no more than a small "insect bite" which goes unnoticed amongst all the others, while this inconspicuous lesion is followed by a long period of latency, with no symptoms, until the person falls down dead at the age of 40 or 50 with a ruptured aneurysm; in fact one endemic focus in Brazil is known as the 'Land of Sudden Death'. Sometimes the course is less dramatic, and, as with Darwin, a chronic illness arises, with signs and symptoms so insidious that the correct diagnosis is often missed.'

ill like cats and mice; almost all chronic illness is surrounded by a penumbra of gloomy imaginings and by worries and fears that may have physical manifestations. I believe that Darwin was organically ill (the case for his having had Chagas' disease is clearly a strong one) but was also the victim of neurosis; and that the neurotic element in his illness may have been caused by the very obscurity of its origins; by his being 'genuinely' ill, that is to say, and having nothing to show for it – surely a great embarrassment to a man whose whole intellectual life was a marshalling and assay of hard evidence. It is a familiar enough story. Ill people suspected of hypochondria or malingering have to pretend to be iller than they really are and may then get taken in by their own deception. They do this to convince others, but Darwin had also to convince himself, for he had no privileged insight into what was wrong with him. The entries in Darwin's notebooks that bear on his health read to me like the writings of a man desperately reassuring himself of the reality of his illness. 'There,' one can imagine his saying, 'I *am* ill, I must be ill; for how otherwise could I feel like this?'

If this interpretation represents any large part of the truth the physicians who inclined to think Darwin a hypochondriac cannot be held blameless, in spite of the fact that the diagnosis of his ailment, if it was indeed Chagas' Disease, was entirely beyond their competence. Even among the tough-minded, the mistaken diagnosis of neurotic illness may cause an extreme exasperation – with symptoms which, of course, serve only to confirm the physician in his diagnosis. But Darwin was a gentle creature who had greatly revered his physician father: to such a man the implied diagnosis of hypochondria would carry special authority and do grave and lasting harm. Perhaps Darwin's physicians should have been more on their guard against an interpretation of his illness that gave him so much less comfort than it gave themselves.[1]

[1] Among the articles that have appeared since this was written are W. D. Foster: *Bull. Hist. Med.* **39**, p. 476, 1965; A. W. Woodruff: *Brit. med. J.* **1**, p. 745, 1965; S. Adler, ibid, p. 1249; *The Lancet* (leading article) **1**, p. 1150, 1965; and I should not have overlooked L. A. Kohn, *Bull. Hist. Med.* **37**, p. 239, 1963.

The Phenomenon of Man

The Phenomenon of Man. By PIERRE TEILHARD DE CHARDIN. With an introduction by Sir Julian Huxley. Collins, London, 1959. 25s.

Everything does not happen continuously at any one moment in the universe. Neither does everything happen everywhere in it.

There are no summits without abysses.

When the end of the world is mentioned, the idea that leaps into our minds is always one of catastrophe.

Life was born and propagates itself on the earth as a solitary pulsation.

In the last analysis the best guarantee that a thing should happen is that it appears to us as vitally necessary.

This little bouquet of aphorisms, each one thought sufficiently important by its author to deserve a paragraph to itself, is taken from Père Teilhard's *The Phenomenon of Man.* It is a book widely held to be of the utmost profundity and significance; it created something like a sensation upon its publication a few years ago in France, and some reviewers hereabouts have called it the Book of the Year – one, the Book of the Century. Yet the greater part of it, I shall show, is nonsense, tricked out with a variety of tedious metaphysical conceits, and its author can be excused of dishonesty only on the grounds that before deceiving others he has taken great pains to deceive himself. *The Phenomenon of Man* cannot be read without a feeling of suffocation, a gasping and flailing around for sense. There is an argument in it, to be sure – a feeble argument, abominably expressed – and this I shall expound in due course; but consider first the style, because it is the style that creates the illusion of content, and which is in some part the cause as well as merely the symptom of Teilhard's alarming apocalyptic seizures.

The Phenomenon of Man stands square in the tradition of *Natur-philosophie*, a philosophical indoor pastime of German origin which does not seem even by accident (though there is a great deal of it) to have contributed anything of permanent value to the store-house of human thought. French is not a language that lends itself naturally to the opaque and ponderous idiom of nature-philosophy, and Teilhard has accordingly resorted to the use of that tipsy, euphoric prose-poetry which is one of the more tiresome mani-festations of the French spirit. It is of the nature of reproduction that progeny should outnumber parents, and of Mendelian here-dity that the inborn endowments of the parents should be variously recombined and reassorted among their offspring, so enlarging the population's candidature for evolutionary change. Teilhard puts the matter thus: it is one of his more lucid passages, and Mr Wall's translation, here as almost everywhere else, cap-tures the spirit and sense of the original.

> Reproduction doubles the mother cell. Thus, by a mechanism which is the inverse of chemical disintegration, *it multiplies without crumbling*. At the same time, however, it transforms what was only intended to be prolonged. Closed in on itself, the living element reaches more or less quickly a state of immobility. It becomes stuck and coagulated in its evolution. Then by the act of reproduction it regains the faculty for inner re-adjustment and consequently takes on a new appearance and direction. The process is one of pluralization in form as well as in number. The elemental ripple of life that emerges from each individual unit does not spread outwards in a monotonous circle formed of individual units exactly like itself. It is diffracted and becomes iridescent, with an indefinite scale of variegated tonalities. The living unit is a centre of irresistible multiplication, and *ipso facto* an equally irresistible focus of diversification.

In no sense other than an utterly trivial one is reproduction the inverse of chemical disintegration. It is a misunderstanding of genetics to suppose that reproduction is only 'intended' to make facsimiles, for parasexual processes of genetical exchange are to be found in the simplest living things. There seems to be some confusion between the versatility of a population and the adaptability of an individual. But errors of fact or judgement of this kind are to be found throughout, and are not my immediate concern; notice instead the use of adjectives of excess (misuse, rather, for genetic diversity is not indefinite nor multiplication irresistible). Teilhard is for ever shouting at us: things or affairs are, in alphabetical order, astounding, colossal, endless, enormous, fantastic, giddy, hyper-, immense, implacable, indefinite, inexhaustible, inextricable, infinite, infinitesimal, innumerable, irresistible, measureless, mega-, monstrous, mysterious, prodigious, relentless, super-, ultra-, unbelievable, unbridled, or unparalleled. When something is described as merely *huge* we feel let down. After this softening-up process we are ready to take delivery of the neologisms: biota, noosphere, hominization, complexification. There is much else in the literary idiom of nature-philosophy: *nothing-buttery*, for example, always part of the minor symptomatology of the bogus. 'Love in all its subtleties is nothing more, and nothing less, than the more or less direct trace marked on the heart of the element by the psychical convergence of the universe upon itself.' 'Man discovers that he is *nothing else than evolution become conscious of itself,*' and evolution is 'nothing else than the continual growth of . . . "psychic" or "radial" energy'. Again, 'the Christogenesis of St Paul and St John is nothing else and nothing less than the extension . . . of that noogenesis in which cosmogenesis . . . culminates'. It would have been a great disappointment to me if Vibration did not somewhere make itself felt, for all scientistic mystics either vibrate in person or find themselves resonant with cosmic vibrations; but I am happy to say that on page 266 Teilhard will be found to do so.

These are trivialities, revealing though they are, and perhaps I make too much of them. The evolutionary origins of consciousness are indeed distant and obscure, and perhaps so trite a thought does need this kind of dressing to make it palatable: 'refracted rearwards along the course of evolution, consciousness displays itself qualitatively as a spectrum of shifting hints whose lower

terms are lost in the night'. (The roman type is mine.) What is
much more serious is the fact that Teilhard habitually and syste-
matically cheats with words. His work, he has assured us, is to be
read, not as a metaphysical system, but 'purely and simply as a
scientific treatise' executed with 'remorseless' or 'inescapable'
logic; yet he uses in metaphor words like energy, tension, force,
impetus, and dimension *as if* they retained the weight and thrust of
their special scientific usages. Consciousness, for example, is a
matter upon which Teilhard has been said to have illuminating
views. For the most part consciousness is treated as a manifestation
of energy, though this does not help us very much because the
word 'energy' is itself debauched; but elsewhere we learn that
consciousness is a dimension, or is something with mass, or is something corpuscular and particulate which can exist in various
degrees of concentration, being sometimes infinitely diffuse. In
his lay capacity Teilhard, a naturalist, practised a comparatively
humble and unexacting kind of science, but he must have known
better than to play such tricks as these. On page 60 we read: 'The
simplest form of protoplasm is already a substance of unheard-of
complexity. This complexity increases in geometrical progression
as we pass from the protozoon higher and higher up the scale of
the metazoa. And so it is for the whole of the remainder always
and everywhere.' Later we are told that the '*nascent* cellular world
shows itself to be already infinitely complex'. This seems to leave
little room for improvement. In any event complexity (a subject
on which Teilhard has a great deal to say) is not measurable in
those scalar quantities to which the concept of a geometrical
progression applies.

In spite of all the obstacles that Teilhard perhaps wisely puts in
our way, it is possible to discern a train of thought in *The Phenomenon of Man*. It is founded upon the belief that the fundamental
process or motion in the entire universe is *evolution*, and evolution
is 'a general condition to which all theories, all hypotheses, all
systems must bow . . . a light illuminating all facts, a curve that all
lines must follow'. This being so, it follows that 'nothing could

ever burst forth as final across the different thresholds successively traversed by evolution . . . which has not already existed in an obscure and primordial way' (again my romans). Nothing is wholly new: there is always some primordium or rudiment or archetype of whatever exists or has existed. Love, for example – 'that is to say, the affinity of being with being' – is to be found in some form throughout the organic world, and even at a 'prodigiously rudimentary level', for if there were no such affinity between atoms when they unite into molecules it would be 'physically impossible for love to appear higher up, with us, in "hominized" form'. But above all consciousness is not new, for this would contradict the evolutionary axiom; on the contrary, we are 'logically forced to assume the existence in rudimentary form . . . of some sort of psyche in every corpuscle', even in molecules; 'by the very fact of the individualization of our planet, a certain mass of elementary consciousness was originally imprisoned in the matter of earth'.

What form does this elementary consciousness take? Scientists have not been able to spot it, for they are shallow superficial fellows, unable to see into the inwardness of things – 'up to now, has science ever troubled to look at the world other than from *without?*' Consciousness is an interiority of matter, an 'inner face that everywhere duplicates the "material" external face, which alone is commonly considered by science'. To grasp the nature of the within of things we must understand that energy is of two kinds: the 'tangential', which is energy as scientists use that word, and a radial energy (a term used interchangeably with spiritual or psychic energy) of which consciousness is treated sometimes as the equivalent, sometimes as the manifestation, and sometimes as the consequence (there is no knowing what Teilhard intends). Radial energy appears to be a measure of, or that which conduces towards, complexity or degree of arrangement; thus 'spiritual energy, by its very nature, increases in "radial" value . . . in step with the increasing chemical complexity of the elements of which it represents the inner lining'. It confers *centricity*, and 'the increase of the synthetic state of matter involves . . . an increase of consciousness'.

We are now therefore in a position to understand what evolution is (is nothing but). Evolution is 'the continual growth of . . . "psychic" or "radial" energy, in the course of duration, beneath and within the mechanical energy I called "tangential"'; evolution, then, is 'an ascent towards consciousness'. It follows that evolution must have a 'precise *orientation* and a privileged *axis*' at the topmost pole of which lies Man, born 'a direct lineal descendant from a total effort of life'.

Let us fill in the intermediate stages. Teilhard, with a penetrating insight that Sir Julian Huxley singles out for special praise, discerns that consciousness in the everyday sense is somehow associated with the possession of nervous systems and brains ('we have every reason to think that in animals too a certain inwardness exists, approximately proportional to the development of their brains'). The direction of evolution must therefore be towards cerebralization, i.e. towards becoming brainier. 'Among the infinite modalities in which the complication of life is dispersed,' he tells us, 'the differentiation of nervous tissue stands out . . . as a significant transformation. *It provides a direction*; and by its consequences *it proves that evolution has a direction*.' All else is equivocal and insignificant; in the process of becoming brainier we find 'the very essence of complexity, of essential metamorphosis'. And if we study the evolution of living things, organic evolution, we shall find that in every one of its lines, except only in those in which it does not occur, evolution is an evolution towards increasing complexity of the nervous system and cerebralization. Plants don't count, to be sure (because 'in the vegetable kingdom we are unable to follow along a nervous system the evolution of a psychism obviously remaining diffuse') and the contemplation of insects provokes a certain shuffling of the feet (p. 153); but primates are 'a phylum of *pure and direct cerebralization*' and among them 'evolution went straight to work on the brain, neglecting everything else'. Here is Teilhard's description of noogenesis, the birth of higher consciousness among the primates, and of the noosphere in which that higher consciousness is deployed:

By the end of the Tertiary era, the psychical temperature in the cellular world had been rising for more than 500 million years. . . . When the anthropoid, so to speak, had been brought 'mentally' to boiling-point some further calories were added. . . . No more was needed for the whole inner equilibrium to be upset. . . . By a tiny 'tangential' increase, the 'radial' was turned back on itself and so to speak took an infinite leap forward. Outwardly, almost nothing in the organs had changed. But in depth, a great revolution had taken place: consciousness was now leaping and boiling in a space of super-sensory relationships and representations. . . .

The analogy, it should be explained, is with the vaporization of water when it is brought to boiling-point, and the image of hot vapour remains when all else is forgotten.

I do not propose to criticize the fatuous argument I have just outlined; here, to expound is to expose. What Teilhard seems to be trying to say is that evolution is often (he says always) accompanied by an increase of orderliness or internal coherence or degree of integration. In what sense is the fertilized egg that develops into an adult human being 'higher' than, say, a bacterial cell? In the sense that it contains richer and more complicated genetical instructions for the execution of those processes that together constitute development. Thus Teilhard's radial, spiritual or psychic energy may be equated to 'information' or 'information content' in the sense that has been made reasonably precise by modern communications engineers. To equate it to consciousness, or to regard degree of consciousness as a measure of information content, is one of the silly little metaphysical conceits I mentioned in an earlier paragraph. Teilhard's belief, enthusiastically shared by Sir Julian Huxley, that evolution flouts or foils the second law of thermodynamics is based on a confusion of thought; and the idea that evolution has a main track or privileged axis is unsupported by scientific evidence.

Teilhard is widely believed to have rejected the modern

Mendelian–Darwinian theory of evolution or to have demon-
strated its inadequacy. Certainly he imports a ghost, the entelechy
or *élan vital* of an earlier terminology, into the Mendelian machine;
but he seems to accept the idea that evolution is probationary and
exploratory and mediated through a selective process, a 'groping ,
a 'billionfold trial and error'; 'far be it from me', he declares, 'to
deny its importance'. Unhappily Teilhard has no grasp of the real
weakness of modern evolutionary theory, namely its lack of a
complete theory of variation, of the origin of *candidature* for
evolution. It is not enough to say that 'mutation' is ultimately the
source of all genetical diversity, for that is merely to give the
phenomenon a name: mutation is so defined. What we want, and
are very slowly beginning to get, is a comprehensive theory of the
forms in which new genetical information comes into being. It
may, as I have hinted elsewhere, turn out to be of the nature of
nucleic acids and the chromosomal apparatus that they tend spontaneously to proffer genetical variants – genetical solutions of the
problem of remaining alive – which are more complex and more
elaborate than the immediate occasion calls for; but to construe
this 'complexification' as a manifestation of consciousness is a
wilful abuse of words.

Teilhard's metaphysical argument begins where the scientific
argument leaves off, and the gist of it is extremely simple. Inasmuch as evolution is the fundamental motion of the entire
universe, an ascent along a privileged and necessary pathway
towards consciousness, so it follows that our present consciousness
must 'culminate forwards in some sort of supreme consciousness'.
In expounding this thesis, Teilhard becomes more and more confused and excited and finally almost hysterical. The Supreme
Consciousness, which apparently assimilates to itself all our personal consciousnesses, is, or is embodied in, 'Omega' or the
Omega-point; in Omega 'the movement of synthesis culminates'.
Now Omega is 'already in existence and operative at the very core
of the thinking mass', so if we have our wits about us we should
at this moment be able to detect Omega as 'some excess of

personal, extra-human energy', the more detailed contemplation of which will disclose the Great Presence. Although already in existence, Omega is added to progressively: 'All round us, one by one, like a continual exhalation, "souls" break away, carrying upwards their incommunicable load of consciousness', and so we end up with 'a harmonized collectivity of consciousnesses equivalent to a sort of super-consciousness'.

Teilhard devotes some little thought to the apparently insuperable problem of how to reconcile the persistence of individual consciousnesses with their assimilation to Omega. But the problem yields to the application of 'remorseless logic'. The individual particles of consciousness do not join up any old how, but only centre to centre, thanks to the mediation of Love; Omega, then, 'in its ultimate principle, can only be a distinct Centre radiating at the core of a system of centres', and the final state of the world is one in which 'unity coincides with a paroxysm of harmonized complexity'. And so our hero escapes from his dire predicament: with one bound, Jack was free.

Although elsewhere Teilhard has dared to write an equation so explicit as 'Evolution = Rise of Consciousness' he does not go so far as to write 'Omega = God'; but in the course of some obscure pious rant he does tell us that God, like Omega, is a 'Centre of centres', and in one place he refers to 'God-Omega'.

How have people come to be taken in by *The Phenomenon of Man?* We must not underestimate the size of the market for works of this kind, for philosophy-fiction. Just as compulsory primary education created a market catered for by cheap dailies and weeklies, so the spread of secondary and latterly of tertiary education has created a large population of people, often with well-developed literary and scholarly tastes, who have been educated far beyond their capacity to undertake analytical thought. It is through their eyes that we must attempt to see the attractions of Teilhard, which I shall jot down in the order in which they come to mind.

1. *The Phenomenon of Man* is anti-scientific in temper (scientists

are shown up as shallow folk skating about on the surface of things), and, as if that were not recommendation enough, it was written by a scientist, a fact which seems to give it particular authority and weight. Laymen firmly believe that scientists are one species of person. They are not to know that the different branches of science require very different aptitudes and degrees of skill for their prosecution. Teilhard practised an intellectually unexacting kind of science in which he achieved a moderate proficiency. He has no grasp of what makes a logical argument or of what makes for proof. He does not even preserve the common decencies of scientific writing, though his book is professedly a scientific treatise.

2. It is written in an all but totally unintelligible style, and this is construed as *prima facie* evidence of profundity. (At present this applies only to works of French authorship; in later Victorian and Edwardian times the same deference was thought due to Germans, with equally little reason.) It is because Teilhard has such wonderful *deep* thoughts that he's so difficult to follow – really it's beyond my poor brain but doesn't that just *show* how profound and important it must be?

3. It declares that Man is in a sorry state, the victim of a 'fundamental anguish of being', a 'malady of space-time', a sickness of 'cosmic gravity'. The Predicament of Man is all the rage now that people have sufficient leisure and are sufficiently well fed to contemplate it, and many a tidy literary reputation has bee nbuilt upon exploiting it; anybody nowadays who dared to suggest that the plight of man might not be wholly desperate would get a sharp rap over the knuckles in any literary weekly. Teilhard not only diagnoses in everyone the fashionable disease but propounds a remedy for it – yet a remedy so obscure and so remote from the possibility of application that it is not likely to deprive any practitioner of a living.

4. *The Phenomenon of Man* was introduced to the English-speaking world by Sir Julian Huxley, who, like myself, finds Teilhard somewhat difficult to follow ('If I understood him aright', p. 16 and again p. 18; 'here his thought is not fully clear to

me', p. 19; etc.). Unlike myself, Sir Julian finds Teilhard in possession of a 'rigorous sense of values', one who 'always endeavoured to think concretely'; he was speculative, to be sure, but his speculation was 'always disciplined by logic'. But then it does not seem to me that Huxley expounds Teilhard's argument; his Introduction does little more than to call attention to parallels between Teilhard's thinking and his own. Chief among these is the cosmic significance attached to a suitably generalized conception of evolution – a conception so diluted or attenuated in the course of being generalized as to cover all events or phenomena that are not immobile in time (pp. 12, 13). In particular, Huxley applauds the, in my opinion, mistaken belief that the so-called 'psychosocial evolution' of mankind and the genetical evolution of living organisms generally are two episodes of a continuous integral process (though separated by a 'critical point', whatever that may mean). Yet for all this Huxley finds it impossible to follow Teilhard 'all the way in his gallant attempt to reconcile the supernatural elements in Christianity with the facts and implications of evolution'. But, bless my soul, this reconciliation is just what Teilhard's book is *about*!

I have read and studied *The Phenomenon of Man* with real distress, even with despair. Instead of wringing our hands over the Human Predicament, we should attend to those parts of it which are wholly remediable, above all to the gullibility which makes it possible for people to be taken in by such a bag of tricks as this. If it were an innocent, passive gullibility it would be excusable; but all too clearly, alas, it is an active willingness to be deceived.

F

The Act of Creation

The author of *Darkness at Noon* must be listened to attentively, no matter what he may choose to write upon. Arthur Koestler is a very clever, knowledgeable and inventive man, and *The Act of Creation*[1] is very clever too, and full of information, and quite wonderfully inventive in the use of words. Many of the points it makes are not likely to be challenged. That wit and creative thought have much in common; that great syntheses may be come upon by logically unmapped pathways; that putting two and two together is an important element in discovery and also, in a certain sense, in making jokes: it has all been said before, of course, and in fewer than 750 pages, though never with such vitality; and anyhow much of it will bear repeating. But as a serious and original work of learning I am sorry to say that, in my opinion, *The Act of Creation* simply won't do. This is not because of its amateurishness, which is more often than not endearing, nor even because of its blunders – they don't affect Koestler's arguments very much one way or another, even when they reveal a deep-seated misunderstanding of, for example, 'Neo-Darwinism', or find expression in fatuous epigrams like 'All automatic functions of the body are patterned by rhythmic pulsations'. I shall try to explain later what I think wrong with Koestler's technical arguments, but let us first of all examine *The Act of Creation* at the level of philosophical *belles lettres*.

As to style, Koestler overdoes it. On one half-page catharsis is described as an 'earthing' of the emotions, the satisfaction of seeing a joke is said to supply 'added voltage to the original charge detonated in laughter', and a smutty story is put at 'the infra-red end of the emotive spectrum'. We aren't quite sure when he intends to be taken literally: for example, what about 'A concept has as many dimensions in semantic space as there are matrices of

[1] Hutchinson (London, 1964).

which it is a member'? This could have been intended to express an exact idea, for 'space' and 'dimension' have generalized technical meanings; but the feeble passage that follows, illustrated by the word 'Madrid', tells us only that the word itself and the city it stands for conjure up all kinds of different associations in his mind.

When I started Koestler's book I hoped he was going to do something for which his knowledge and sympathies and writer's insight should give him unequalled qualifications: that he would give us a first ethology of scientific activity, and so help to make the scientist intelligible to others and to himself. Unhappily, there are passages in *The Act of Creation* which convince me that he has no real grasp of how scientists go about their work. Consider the depths of misunderstanding revealed by Koestler's aloof and snobbish remarks about the 'unseemly haste' with which some scientists publish their discoveries. Koestler's historical hobnob-bings with men of genius seem to have made him forget the fact that, in science, what X misses today Y will surely hit upon tomorrow (or maybe the day after tomorrow). Much of a scientist's pride and sense of accomplishment turns therefore upon being the *first* to do something – upon being the man who did actually speed up or redirect the flow of thought and growth of understanding. There is no spiritual copyright in scientific discoveries, unless they should happen to be quite mistaken. Only in making a blunder does a scientist do something which, conceivably, no one else might ever do again. Artists are not troubled by matters of priority, but Wagner would cer-tainly not have spent twenty years on *The Ring* if he had thought it at all possible for someone else to nip in ahead of him with *Götterdämmerung*.

Like other amateurs, Koestler finds it difficult to understand why scientists seem so often to shirk the study of really fundamen-tal or challenging problems. With Robert Graves he regrets the absence of 'intense research' upon variations in the – ah – 'emotive potentials of the sense modalities'. He wonders why 'the genetics of behaviour' should still be 'uncharted territory' and asks whether this may not be because the framework of Neo-Dar-winism is too rickety to support an inquiry. The real reason is so much simpler: the problem is very, very difficult. Goodness knows how it is to be got at. It may be outflanked or it may yield to attrition, but probably not to direct assault. No scientist is admired for failing in the attempt to solve problems that lie beyond his competence. The most he can hope for is the kindly contempt earned by the Utopian politician. If politics is the art of the possible, research is surely the art of the soluble. Both are immensely practical-minded affairs.

Although much of Koestler's book has to do with explanation, he seems to pay little attention to the narrowly scientific usages of the concept. Some of the 'explanations' he quotes with approval[1] are simply analgesic pills which dull the aches of incomprehension without going to their causes. The kind of explanation the scientist spends most of his time thinking up and testing – the hypothesis which enfolds the matters to be explained among its logical consequences – gets little attention. Instead we are told that there are all kinds of explanations and many degrees of understanding, starting with the '*unconscious* understanding mediated by the dream'.

Dreams bring out the worst in Koestler. Dreaming is a 'sliding back towards the pulsating darkness, one and undivided, of which we were part before our separate egos were formed'. No wonder, then, that the understanding it conveys is of the unconscious kind. 'There is no need to emphasize, in this century of Freud and Jung, that the logic of the dream . . . derives from the magic type of causation found in primitive societies and the fantasies of child-hood.' But those who enjoy slopping around in the amniotic fluid should pause for a moment to entertain (perhaps only unconsciously in the first instance) the idea that the content of dreams may be totally devoid of 'meaning'. There should be no need to emphasize, in this century of radio sets and electronic devices, that many dreams may be assemblages of

---

[1] For example, his account on page 452 of C. M. Child's explanations of 'physiological isolation'.

thought-elements that convey no information whatsoever: that they may just be *noise*.[1]

Koestler's theory of the creative act is set out in Book One as a special theory comprehended within a General Theory that occupies Book Two. In Book One he defines two special notions, 'matrix' and 'code'. A code is a system of rules of process or performance and a matrix is 'any ability, habit, or skill, any pattern of ordered behaviour' governed by a code. In particular, a matrix of thought is, or can for variety's sake be described as, a 'frame of reference', an 'associative context', a 'type of logic', or a 'universe of discourse'. Behind every act of creation lies a binary association (bisociation) of matrices: in that which provokes laughter they collide, in a new intellectual synthesis they fuse, and in an aesthetic experience they confront each other or are juxtaposed. These three degrees of experience form a continuum, and may indeed grow out of the bisociation of the very same matrices. Thus the exuberant, explosive, tension-relieving delight *(Eureka!)* and the long after-glow ('the oceanic feeling') excited by an intellectual synthesis have their counterparts in laughter and in the sense of satisfaction at seeing a joke.

Koestler is far too intelligent a man not to realize that his account of creative activity is full of difficulties, but though he mentions the contexts in which some of them arise, he does not direct attention to them explicitly or make any attempt to work them out. Among them, and in no special order, are: *(a)* just how does an explanation which later proves false (as most do: and none, he admits, is proved true) give rise to just the same feelings of joy and exaltation as one which later stands up to challenge? What went wrong: didn't the matrices fuse, or were they the wrong kind of matrix, or what? *(b)* The source of most joy in science lies not so much in devising an explanation as in getting the results of an experiment which upholds it. *(c)* Some awkward problems are raised by the fact that the chap who *sees* a joke splits his sides laughing as well, maybe, as the chap who makes it; but the chap

---

[1] 'Noise' as the word is used by communication theorists: I have borrowed this thought from an impromptu of Professor A. L. Hodgkin's.

who 'sees' or is apprised of an intellectual synthesis does not share in the tension-relieving, explosive joy of discovery. Likewise the joy of artistic creation 'travels' in some sense in which the joy of intellectual synthesis does not, and this difference between them seems to me to outweigh their similarities. *(d)* It follows from Koestler's scheme that an intellectual synthesis, upon being proved false, should at once become a huge joke, especially to the person who devised it, for it must have rested upon the kind of bisociative act that underlies the comical. However, we are not amused. *(e)* The sense of comfort an explanation may give rise to has nothing to do with bringing about or even witnessing a 'fusion of matrices': laymen get it not so much from knowing an explanation as from knowing that an explanation is known.

I should mention, because Koestler does not, that the so-called 'hypothetico-deductive' interpretation of the scientific process copes perfectly with difficulties *(a)* and *(b)*, and that in it *(c)*, *(d)* and *(e)* do not arise. Devising a hypothesis is a 'creative act' in the sense that it is the invention of a possible world, or a possible fragment of the world; experiments are then done to find out whether or not that imagined world is, to a good enough approximation, the real one. As Koestler conceives it, the act of creation is not, in the usual sense, creative at all; as he says, it merely 'uncovers, selects, reshuffles, combines, synthesises already existing facts, ideas, faculties, skills'.

Koestler's psychological thought, though not confessedly 'introspective', is in the style of the nineteenth century – a point delicately made, as I read it, by Sir Cyril Burt in his foreword. Koestler nags away at Behaviourism, which he describes as 'the dominant school in contemporary psychology', though later he says of J. B. Watson's textbook that 'few students today remember its contents, or even its basic postulates.' For all its crudities Behaviourism, conceived as a methodology rather than as a psychological system, taught psychology with brutal emphasis that 'the dog is whining' and 'the dog is sad' are statements of altogether different empirical standing, and heaven help psychology if it ever again overlooks the distinction.

I was dreading the moment in my reading of Koestler when I was to be told that sexual reproduction, 'the bisociation of two genetic codes', was 'the basic model of the creative act'. Koestler's Book Two contains the general theory which comprehends the special theory I have just outlined. It is full of rather old-fashioned biology (but what fun to read again of axial gradients!), and the person who has to be told on page 57 that the sympathetic nervous system has nothing to do with the emotion of sympathy will not make much of it. The argument runs thus. The system of nature is a hierarchical system of elements, sub-elements, sub-sub-elements, etc., each enjoying a certain wholeness and autonomy, but each also subordinate to the element above it in the hierarchy. The structure of army command is one example of such a hierarchy (a company has some autonomy but is subordinate to the battalion) and in the living world the hierarchy of organism, organ, cell, cell part, etc. is another. At each level we find a certain wholeness and a certain partness.

Koestler declares that at each level of the hierarchy 'homologous' principles operate, with the consequence that any phenomenon at one level must have its homologue or formal counterpart at each other level. In particular, we shall find 'mental equivalents' of what goes on at lower levels in the hierarchy, and conversely, since we can go either up or down the ladder of correspondences, physical equivalents of what goes on in the mind. The 'creative stress' of the artist or scientist corresponds to the 'general alarm reaction' of the injured animal, and dreaming is the mental equivalent of the regenerative processes that make good wear and tear. Embryonic development has a certain self-assertive quality, and so have 'perceptual matrices'. A not yet verbalized analogy corresponds to an organ rudiment in the early embryo, and rhythm and rhyme, assonance and pun, are 'vestigial echoes' of the 'primitive pulsations of living matter'.

No metaphors these: 'they have solid roots in the earth'; but to my ears they sound silly, and I believe them to be as silly as they sound. Disregarding the rights and wrongs of building hierarchies out of non-homogeneous elements (though in fact it won't do to mix up perceptual matrices with adrenal glands, embryos, jokes and rhymes), the correspondences Koestler makes so much of are of the purely formal and abstract kind that can be expressed without any regard to their empirical content. Even if the networks of relationships holding at each level of the hierarchy were isomorphic, there would be no necessary affinity between the things so related. The correspondences which Koestler urges us to believe in are harmless enough, but arguments of this kind can be mischievous (e.g. a case for totalitarian government can be built upon an unsound analogy between Organism and State).

Koestler makes a good point when he says that during the past hundred years or less scientists have felt themselves under a professional obligation to write in a dry, cold, pulseless way; to be, in short, boring. (It is part of the heritage of inductivism.) *The Act of Creation* is so full of vitality that it creates around it an aura of good-to-be-alive, and though Koestler regards himself as the author of a new and important general psychological theory, I am delighted that, in writing a 'popular' work, he has in a sense appealed to the general public over the heads of the profession. But certain rules of scientific manners must be observed no matter what form the account of a scientific theory may take. One must mention (if only to dismiss with contempt) other, alternative explanations of the matters one is dealing with; and one must discuss (if only to prove them groundless) some of the objections that are likely to be raised against one's theories by the ignorant or ill-disposed. Koestler seems to have no adequate grasp of the importance of *criticism* in science – above all of self-criticism, for most of a scientist's wounds are self-inflicted. Nor can I remember in his book any passage suggesting observations or experiments which might qualify or refine his ideas. He quotes with approval one 'laconic pronouncement' of Dirac's which must have made sense in context but which otherwise sounds just naughty: 'It is more important to have beauty in one's equations than to have them fit experiment.' The high inspirational origin of a theory is no guarantee of its trustworthiness, and Koestler should avoid giving the impression that he thinks it is. No belief could bring science more quickly to ruin.

Mr Koestler replied thus (New Statesman, 19 June 1964):

Sɪʀ – Allow me to answer some of the points raised in Professor Medawar's review of my recent book.

1. Medawar writes: 'There are passages in *The Act of Creation* which convince me that he has no real grasp of how scientists go about their work. Consider the depths of misunderstanding revealed by Koestler's aloof and snobbish remarks about the "unseemly haste" with which some scientists publish their discoveries. Koestler's historical hobnobbings with men of genius seem to have made him forget the fact that, in science, what X misses today Y will surely hit upon tomorrow . . .'

The snobbish remarks to which this passage refers read as follows: 'In 1922, Ogburn and Thomas published some 150 examples of discoveries and inventions which were made independently by several persons; and, more recently, Merton came to the seemingly paradoxical conclusion that "the pattern of independent multiple discoveries in science is . . . the dominant pattern rather than a subsidiary one." He quotes as an example Lord Kelvin, whose published papers contain "at least 32 discoveries of his own which he subsequently found had also been made by others. . ." The endless priority disputes which have poisoned the supposedly serene atmosphere of scientific research throughout the ages, and the unseemly haste of many scientists to establish priority by rushing into print – or, at least, depositing manuscripts in sealed envelopes with some learned society – point in the same direction. Some – among them Galileo and Hooke – even went to the length of publishing half-completed discoveries in the form of anagrams, to ensure priority without letting rivals in on the idea.'

2. Medawar seems to object to my quite unoriginal contention that unconscious processes in the dream and in the hypnagogic state between dreaming and awakening often play a decisive part in scientific discovery. At least this seems to be the

meaning behind the heavy veils of irony in the passage: 'Dreams bring out the worst in Koestler . . . Those who enjoy slopping around in the amniotic fluid should pause for a moment to entertain (perhaps only unconsciously in the first instance) the idea that the content of dreams may be totally devoid of "meaning" . . . that many dreams may be assemblages of thought elements that convey no information whatsoever.'

No doubt most dreams are self-addressed messages whose information-content is purely private and 'meaningless' to others. But equally undeniable is the fact – which Medawar chooses to pass in silence – that dreams, hypnagogic images and other forms of unconscious intuitions proved decisive in the discoveries of dozens of scientists and mathematicians whose testimonies I quoted – among them Ampère, Gauss, Kékulé, Leibnitz, Poincaré, Fechner, Otto Loewi, Planck, Einstein, to mention only a few.

3. Medawar raises five objections, numbered *(a)* to *(e)*, against the theory I proposed. To save space, let me refer to the passages in the book in which the answers to these objections can be found. *(a)* Bk. One, Chap. IX, p. 212 et seq. *(b)* The 'joy' in 'devising an explanation' and the satisfaction derived from its empirical confirmation enter at different stages and must not be confused. *(c)* Bk. One, Chap. IV, p. 87 et seq. and Chap. XI, p. 255 et seq. *(d)* 'It follows from Koestler's scheme, etc'. The answer is, it does not. *(e)* See Bk. One, Chap. XVII, pp. 325–31.

4. Medawar accuses me of quoting out of context 'one laconic pronouncement of Dirac's.' The single sentence which Medawar requotes is on p. 329. The full context, which Medawar overlooked, is to be found on pp. 245–6. If he had no time to read through the book, he should at least have looked at the index.

5. Medawar accuses me of contradicting myself: 'Koestler nags away at Behaviourism, which he describes as "the dominant school in contemporary psychology", though later he says of J. B. Watson's textbook that "few students today remember its contents or even its basic postulates".' This I said

on p. 558. On p. 559 I continued: 'Although the cruder absurdities of Watsonian behaviourism are forgotten, it had laid the foundations on which the later, more refined behaviouristic systems [of Guthrie, Hull and Skinner] were built; the dominant trend in American and Russian psychology in the generation that followed had a distinctly Pavlov-Watsonian flavour. The methods became more sophisticated, but the philosophy behind them remained the same.'

The rest consists of ironic innuendo and *ex cathedra* pronouncements. 'Certain rules of scientific manners must be observed,' Professor Medawar informs us. I wish he had lived up to his precept.

I in turn answered:

I should like to take Mr Koestler's points one by one.

1. *Priority.* A scientist's sense of concern about matters of priority may not be creditable, but only prigs deny its existence, and the fact that it does exist points towards something distinctive in the act of creation as it occurs in science. It is not good enough to brush it aside with clichés ('unseemly haste', 'rushing into print') or to pour scorn on its extremer manifestations. I think my own interpretation was the right one – priority in science gives moral possession – but Koestler seems not to realize that there is anything to interpret. As to simultaneous discovery: it was the slight air of wonderment about it in the very passage Koestler now quotes which made me ask if he realized the consequences of the relationship, peculiar to science, between X and Y. Simultaneous discovery is utterly commonplace, and it was only the rarity of scientists, not the inherent improbability of the phenomenon, that made it remarkable in the past. Scientists on the same road may be expected to arrive at the same destination, often not far apart. Romantics like Koestler don't like to admit this, because it seems to them to derogate from the authority of genius. Thus of Newton and Leibnitz, equal first with the differential calculus, Koestler says 'the greatness of this accomplishment is hardly diminished by

the fact that two among millions, instead of one among millions, had the exceptional genius to do it.' But millions weren't *trying* for the calculus. If they had been, hundreds would have got it. Very simple-minded people think that if Newton had died prematurely we should still be at our wits' end to account for the fall of apples. Is there not just a trace of this *naïveté* in Koestler?

2. *Dreams.* Koestler quite misses the point, and what he says is a good example of how stubbornly the mind may deny entry to the unfamiliar (*The Act of Creation*, p. 216). I did not suggest that dreams conveyed private messages whose import was known only to the dreamer. My proposal, as unoriginal as Koestler's, was that dreams are not messages at all. It is naughty of Koestler to lump together 'dreams, hypnagogic images and other forms of unconscious intuitions', as if my misgivings about dreams extended to all other unscripted activities of the mind. They don't: if we are to brandish texts at each other, I will cite an article in the *TLS* (25 October 1963) in which I speak up for inspiration and against the idea that discovery can be logically mechanized.

3. *Objections (a), (c)* and *(e)*. Koestler would not have drawn special attention to these passages in his book unless he really believed them to hold the answers to my, as I think, damaging objections to his theory. Now, on rereading him, I feel convinced that he simply doesn't understand the *kind* of intellectual performance that is expected of someone who propounds or defends a scientific or philosophic theory. But now we have both had our say and it can all go out to arbitration. *(b)* We agree, then: but why here no citations of the passages in his book in which he says so? *(d)* I'm so sorry, but I still think it follows from Koestler's scheme that an intellectual synthesis, upon being proved false, should appear funny. (If I were parodying Koestler I should describe it as a joke played by Nature of which we were very slow to see the point.) What's more, the converse also follows, that a great intellectual synthesis wrongly believed false will be thought hugely comical from the outset. Koestler seems to think so too: 'Until the

seventeenth century the Copernican hypothesis of the earth's motion was considered as obviously incompatible with common-sense experience; it was accordingly treated as a huge joke . . . The history of science abounds with examples of discoveries greeted with howls of laughter because they seemed to be a marriage of incompatibles . . .' (pp. 94–5). But as I say, in real life the refutation of a hypothesis can be deeply upsetting. If there are howls they are not of laughter.

4. Dirac's allegedly 'laconic pronouncement' ("It is more important to have beauty in one's equations than to have them fit experiment") is in reality a very diffident expression of opinion whose context I was wrong to overlook. But it was rash of Koestler to draw my attention to it, for what Dirac goes on to say is: 'It seems that if one is working from the point of view of getting beauty in one's equation, *and if one has really a sound insight,* one is on a sure line of progress.' The italics are mine, but Koestler is welcome to them.

5. *Behaviourism.* No, I didn't say Koestler contradicted himself, though I do find his love-hate relationship with modern experimental psychology extremely tiresome. Nor do I think he has quite got my point, which was that even if behaviourism were dead as a system it is still very much alive as a methodology.

Koestler must have had some doubts about the wisdom and taste of his final sentences, and I suppose it will only make matters worse if I say I forgive him. I will not, however, forgive him for hinting that I didn't read his book, nor for the fact that I had to spend hours and hours and hours in doing so.

A Biological Retrospect

The title of my presidential address, you will have discerned, is 'A Biological Retrospect', and on the whole it has not been well received. 'Why a biological *retrospect*?', I have been asked; would it not be more in keeping with the spirit of the occasion if I were to speak of the future of biology rather than of its past? It would indeed be, if only it were possible, but unfortunately it is not. What we want to know about the science of the future is the content and character of future scientific theories and ideas. Unfortunately, it is impossible to predict new ideas – the ideas people are going to have in ten years' or in ten minutes' time – and we are caught in a logical paradox the moment we try to do so. For to predict an idea is to have an idea, and if we have an idea it can no longer be the subject of a prediction. Try completing the sentence 'I predict that at the next meeting of the British Association someone will propound the following new theory of the relationships of elementary particles, *namely* . . .' If I complete the sentence, the theory will not be new next year; if I fail, then I am not making a prediction.

Most people feel more confident in denying that certain things will come to pass than in declaring that they can happen or surely will happen. Many a golden opportunity to remain silent has been squandered by anti-prophets who do not realize that the grounds for declaring something impossible or inconceivable may be undermined by new ideas that cannot be foreseen. Here is an instructive passage from the philosophic writings of a great British physiologist, J. S. Haldane (father of J.B.S.) It comes from *The Philosophy of a Biologist* of 1931, and its subject is the nature of memory in a very general sense that includes 'genetic memory' – for example, the faculty or endowment which ensures that a frog's egg develops into a frog and, indeed, into a particular kind of frog.

Haldane is very critical of the theories of memory propounded by Ewald Hering and Richard Semon, who

> assume that memory in general is dependent on proto-plasmic "engrams", and that germ-cells are furnished with a system of engrams, functioning as guide-posts to all the normal stages of development.

('Engrams', I should explain, are more or less permanent physical memory traces or memory imprints that are thought to act as directive agencies in development.[1]) 'This theory', Haldane goes on to say,

> has quite evidently all the defects of other attempts at mechanistic explanations of development. How such an amazingly complicated system of signposts could function by any physico-chemical process or reproduce itself indefinitely often is inconceivable.[2]

What Haldane found himself unable even to conceive is today a commonplace. Only twelve years after the publication of the passage I quote, Avery and his colleagues had determined the class of chemical compound to which genetic engrams belong. In the meantime our entire conception of 'the gene' has undergone a revolution. Genes are not, as at one time or another people have thought them, samples or models; they are not enzymes or hormones or prosthetic groups or catalysts or, in the ordinary sense, agents of any kind. Genes are *messages*. I think Kalmus[3] was the first to use this form of words, but the idea that a chromosome is a molecular code script containing a specification of development is Schrödinger's.[4]

[1] See *The Mneme* by R. Semon, London, 1921, particularly pp. 24, 113, 180, 211; and Hering's paper 'On Memory, a Universal Attribute of organized Matter' in *Alm. Akad. Wiss. Wien.*, **20**, 253 (1870).

[2] *The Philosophy of a Biologist*, p. 162 (London, 1931).

[3] 'A Cybernetical Aspect of Genetics', *J. Hered.*, **41**, 19 (1950).

[4] Schrödinger, E. *What is Life?*, especially pp. 19-20, 61-62, 68 (Cambridge, 1944). For Weismann's far-sighted views on the matter, see *The Architecture of Matter* by Toulmin, S., and Goodfield, J. (London, 1962).

My purpose in this address is to identify some of the great conceptual advances that have taken place during the past twenty-five or thirty years on four different planes of biological analysis. Working biologists, I should explain, tend nowadays to classify themselves less by 'subjects' than by the analytical levels at which they work – a horizontal classification where the older was vertical. So we have molecular biologists, whose ambition is to interpret biological performances explicitly in terms of molecular structure; we have cellular biologists, biologists who work at the level of whole organisms (the domain of classical physiology), and biologists who study communities or societies of organisms. We can discern each of these four strata within each 'subject' of the traditional, that is, the vertical, classification. There are molecular and cellular geneticists, geneticists in Mendel's sense, and population geneticists. So also in endocrinology or immunology: each is now studied at the molecular and cellular level as well as at the level of whole organisms. They abut into the population level, too: we study the effects of crowding and fighting on the adrenals and so indirectly on reproductive performance, and we study the epidemiological consequences of natural or artificial immunization and the evolutionary consequences of epidemics. I have noticed that a biologist's interests and understanding, and also, in a curious way, his loyalties, tend to spread horizontally, along strata, rather than up and down. Our instinct is to try to master what belongs to our chosen plane of analysis and to leave to others the research that belongs above that level or below. An ecologist in the modern style, a man working to understand the agencies that govern the structure of natural populations in space and time, needs much more than a knowledge of natural history and a map. He must have a good understanding of population genetics and population dynamics generally, and certainly of animal behaviour; more than that, he must grasp climatic physiology and have a feeling for whatever may concern him among the other conventional disciplines in biology (I have already mentioned immunology and endocrinology). There is no compelling reason why he should be able to talk with relaxed fluency about messenger-RNA, and it is not essential that he should ever have heard of it – though an unreasonable feeling that he 'ought' to know something about it is more likely to be found in a good ecologist than in an indifferent one.

I shall now take one example from each of these four planes of biological analysis and try to show how our ideas have changed since the last Cambridge meeting of the British Association in 1938 – a period that corresponds roughly with my own professional lifetime.

POPULATION GENETICS AND EVOLUTION THEORY

Biologists of my generation were still brought up in what I call the 'dynastic' concept of evolution. The course of evolution was unfolded to us in the form of pedigrees or family trees, and we used the old language of universals in speaking of the evolution of *the* dogfish, *the* horse, *the* elephant and, needless to say, of Man.

The dynastic conception coloured our thoughts long after the revival of Darwinism had made it altogether inappropriate. By the 'revival of Darwinism' I mean the reformulation of Darwinism in the language of Mendelian genetics – the work, as we all so very well know, of Fisher, J. B. S. Haldane, Wright, Norton and, in a rather qualified sense, of Lotka and Volterra. The subject of evolutionary change, we now learned, was a population, not a lineage or pedigree: evolution was a systematic secular change in the genetical structure of a population, and natural selection was overwhelmingly its most important agent. But to those brought up in the dynastic style of thinking about evolution it seemed only natural to suppose that the outcome of an evolutionary episode was the devising of a new genotype – of that new genetical formula which conferred the greatest degree of adaptedness in the prevailing circumstances. This improved genetic formula – a new solution of the problem of remaining alive in a hostile environment – would be shared by the great majority of the members of the population, and would be stable except in so far as it might be modified by further evolution. The members of the population were predominantly uniform and homozygous in genetic make-up, and, to whatever degree they were so, would necessarily breed true. Genetic diversity was maintained by an undercurrent of mutation, but most mutants upset the hard-won formula for adaptedness and natural selection forced them into recessive expression, where they could do little harm. When evolution was not in progress natural selection made on the whole for uniformity. Polymorphism, the occurrence of a stable pattern of genetic inequality, was recognized as an interesting but somewhat unusual phenomenon, each example of which required a special explanation, that is, an explanation peculiar to itself.

These ideas have now been superseded, mainly through the empirical discovery that natural outbreeding populations are highly diverse. Chemical polymorphism (allotypy[1]) is found wherever it is looked for intently enough by methods competent to reveal it. The molecular variants known in human blood alone provide combinations that far outnumber the human race – variants of haemoglobin, non-haemoglobin proteins, and red-cell enzymes; of red-cell antigens and white-cell antigens; and of haptoglobins, transferrins and immunoglobulins. Today it is no longer possible to think of the evolutionary process as the formulation of a new genotype or the inauguration of a new type of organism enjoying the possession of that formula. The *product* of evolution is itself a population – a population with a certain newly devised and well adapted pattern of genetic *in*equality. This pattern of genetic differentiation is shaped and actively maintained by selective forces: it is the population as a whole that breeds true, not its individual members. We can no longer draw a distinction between an active process of evolution and a more or less stationary end-product: evolution is constantly in progress, and the genetical structure of a population is actively, that is dynamically, sustained.

These newer ideas have important practical consequences. The older outlook was embodied in that older, almost immemorial ambition of the livestock breeder, to produce by artificial selection

---

[1] A term coined by J. Oudin to describe gamma-globulin variants: it might well be generalized to include all molecular polymorphism.

a true breeding stock with uniform, and uniformly desirable characteristics; and this was also the ambition – sometimes kindly, but always mistaken – of old-fashioned 'positive' eugenics. It now seems doubtful if, with free-living and naturally out-breeding organisms, such a goal can ever be achieved. Modern stock-breeders tend to adopt a very nicely calculated regimen of cross-breeding which, abandoning the goal of a single self-perpetuating stock, achieves a uniform marketable product of hybrid composition. The genetical theory underlying this scheme of breeding embodies, and was indeed partly responsible for, the newer ideas of population structure I have just outlined.

I cannot predict what new ideas will illuminate the theory of evolution in future, but it is not difficult to guess the contexts of thought in which they are likely to appear. The main weakness of modern evolutionary theory is its lack of a fully worked out theory of variation, that is, of *candidature* for evolution, of the forms in which genetic variants are proffered for selection. We have therefore no convincing account of evolutionary progress – of the otherwise inexplicable tendency of organisms to adopt ever more complicated solutions of the problems of remaining alive. This is a 'molecular' problem, in the newer biological usage of that word, because its working out depends on a deeper under-standing of how the physicochemical properties and behaviour of chromosomes and nucleoproteins generally qualify them to enrich the candidature for evolution; and this reflection is my cue to turn to conceptual advances in biology at the molecular level.

THE PHYSICAL BASIS OF LIFE

In the early 1930's no one knew what to make of the nucleic acids. Bawden and Pirie had not yet shown that nucleic acid was an integral part of the structure of tobacco mosaic virus, and we were still a decade from the astonishing discovery by Avery and his colleagues, in the Rockefeller Institute, that the agent responsible for pneumococcal transformations was a deoxyribonucleic acid.

Since there was nothing very much to say about nucleic acids you may well wonder what everybody *did* talk about. One topic of conversation was the crystallization of enzymes. Summer had crystallized urease in 1926 and Northrop pepsin in 1930; soon Stanley would crystallize tobacco mosaic virus, at that time still thought to be a pure protein. But the most exciting and, as it seemed to us, portentous discoveries were those of W. T. Astbury, whose X-ray diffraction pictures of silk fibroin and hair and feather keratins had revealed an essentially crystalline orderliness in ordinary biological structures. For some purposes, however, X-ray analysis was too powerful. The occasion called for resolving powers between those of the optical microscope and the X-ray tube, and this need was fulfilled by electron microscopy. I saw my first electron-photomicrograph in *Nature* in 1933; its resolving power was then one micron.

Electron microscopy has shown that cells contain sheets, tubes, bags and, indeed, micro-organs – real anatomical structures in the sense that they have firm and definite shapes and look as if only their size prevented our picking them up and handling them. Moreover, there is no dividing line between structures in the molecular and in the anatomical sense: macromolecules have structures in a sense intelligible to the anatomist and small anatomical structures are molecular in a sense intelligible to the chemist. (Intelligible *now*, I should add: as Pirie[1] has told us, the idea that molecules have literally, that is, spatially, a structure was resisted by orthodox chemists, and the credentials of molecules with weights above 5,000 were long in doubt.) In short, the orderliness of cells is a structural or crystalline orderliness – a 'solid' orderliness, indeed, for 'the so-called amorphous solids are either not really amorphous or not really solid'.

This newer conception represents a genuine upheaval of biological thought, and it marks the disappearance of what may be called the *colloidal* conception of vital organization, itself a sophisticated variant of the older doctrine of 'protoplasm'. The idea of protoplasm as a fragile colloidal slime, a sort of biological ether permeating otherwise inanimate structures, was already obsolete

[1] Pirie, N. W., 'Patterns of Assumption about Large Molecules', *Arch. Biochem. Biophys.*, Supp. 1, 21 (1962).

in the 'thirties; even then no one could profess to be studying 'protoplasm' without being thought facetious or slightly mad. But we still clung to the colloidal conception in its more sophisticated versions, which allowed for heterogeneity and for the existence of liquid crystalline states, and it was still possible to applaud Hopkins's famous aphorism from the British Association meeting of 1913, that the life of the cell is 'the expression of a particular dynamic equilibrium in a polyphasic system'. For inadequate though the colloidal conception was seen to be, there was nothing to take its place. Peters's idea of the existence of a 'cytoskeleton' to account for the orderly unfolding of cellular metabolism in time and place now seems wonderfully prescient, but there was precious little direct evidence for the existence of anything of the kind, and much that seemed incompatible with it.

The substitution of the structural for the colloidal conception of 'the physical basis of life' was one of the great revolutions of modern biology; but it was a quiet revolution, for no one opposed it, and for that reason, I suppose, no one thought to read a funeral oration over protoplasm itself.

CELLULAR DIFFERENTIATION IN EMBRYONIC DEVELOPMENT

Embryology is in some ways a model science. It has always been distinguished by the exactitude, even punctilio, of its anatomical descriptions. An experiment by one of the grand masters of embryology could be made the text of a discourse on scientific method. But something is wrong; or has been wrong. There is no *theory* of development, in the sense in which Mendelism is a theory that accounts for the results of breeding experiments. There has therefore been little sense of progression or timeliness about embryological research. Of many papers delivered at embryological meetings, however good they may be in themselves – in themselves they are sometimes marvels of analysis, and complete and satisfying within their own limits – one too often feels that they might have been delivered five years beforehand without making anyone much the wiser, or deferred for five years without making anyone conscious of great loss.

It was not always so. In the 1930's experimental embryology had much the same appeal as molecular biology has today: students felt it to be the most rapidly advancing front of biological research. This was partly due to the work of Vogt, who had shown that the mobilization and deployment of cellular envelopes, tubes and sheets was the fundamental stratagem of early vertebrate development (thus re-laying the foundations of comparative vertebrate embryology); but it was mainly due to the 'organizer theory' of Hans Spemann, the theory that differentiation in development is the outcome of an orderly sequence of specific inductive stimuli. The underlying assumption of the theory (though not then so expressed) was that we should look to the chemical properties of the inductive agent to find out why the amino-acid sequence of one enzyme or organ-specific protein should differ from the amino-acid sequence of another. The reactive capabilities of the responding tissue were emphasized repeatedly, but only at a theoretical level, for 'competence' did not lend itself to experimental analysis, and the centre of gravity of actual research lay in the chemical definition of inductive agents.

Wise after the event, we can now see that embryology simply did not have, and could not have created, the background of genetical reasoning which would have made it possible to formulate a theory of development. It is not now generally believed that a stimulus external to the system on which it acts can specify the primary structure of a protein, that is, convey instructions that amino-acids shall be assembled in a given order. The 'instructive' stimulus has gone the way of the philosopher's stone, an agent dimly akin to it in certain ways. Embryonic development at the level of molecular differentiation must therefore be an unfolding of pre-existing capabilities, an acting-out of genetically encoded instructions; the inductive stimulus is the agent that selects or activates one set of instructions rather than another. It is just possible to see how something of the kind happens in the induction of adaptive enzymes in bacteria – a phenomenon of which the older description, the 'training' of bacteria, reminds us that it too, at one time, was thought to be 'instructive' in nature. All this applies only to biological order at the level of the amino-acid sequences of proteins or the nucleotide sequences of nucleic acids. Nothing is yet known about the genetic specification of order at levels above the molecular level.

The function performed by the hierarchy of inductive stimuli as it occurs in vertebrate development is to determine the specificities of time and place: it is an inductive stimulus which determines that a lens shall form here and not there, now and not then. As I see it, it is the inductive process that allows vertebrate eggs and embryos before gastrulation to indulge in the prodigious range of adaptive radiation to be seen in germs as disparate as a dogfish's egg and a human being's – a case I have argued elsewhere and need not go over here again.[1]

BIOLOGY OF THE ORGANISM: ANIMAL BEHAVIOUR

If experimental embryology was the subject that seemed most exciting to students of the thirties, that most nearly on the threshold of a grand revelation, the study of animal behaviour (in the sense in which we now tend to use the word 'ethology'), seemed just as clearly the most frustrating and unrewarding. Twenty years later it was the other way about: embryology had lost much of its fascination and many of the ablest students were recruited into research on behaviour instead. What had happened in the meantime?

In the early 1930's we had one new behavioural concept to ponder on: the idea that an animal might in some way apprehend a sensory pattern or a behavioural situation as a whole and not by a piecing together of its sensory or motor parts. That was the lesson of *Gestalt* theory. We had also learnt finally, and I hope for ever, the methodological lesson of behaviourism: that statements about what an animal feels or is conscious of, and what its motives are, belong to an altogether different class from statements about what it does and how it acts. I say the 'methodological' lesson of

[1] *J. Embryol. Exp. Morph.*, **2**, 172 (1954).

behaviourism, because that word also stands for a certain psychological theory, namely, that the phenomenology of behaviour is the whole of behaviour – a theory of which I shall only say that, in my opinion, it is not nearly as silly as it sounds. Even the methodology of behaviourism seemed cruelly austere to a generation not yet weaned from the doctrine of privileged insight through introspection. But what comparable revolution of thought ushered in the study of animal behaviour in the style of Lorenz and Tinbergen and led to the foundation of flourishing schools of behaviour in Oxford, here in Cambridge, and throughout the world?

I believe the following extremely simple answer to be the right one. In the 'thirties it did not seem to us that there *was* any way of studying behaviour 'scientifically' except through some kind of experimental intervention – except by confronting the subject of our observations with a 'situation' or with a nicely contrived stimulus and then recording what the animal did. The situation would then be varied in some way that seemed appropriate, whereupon the animal's behaviour would also vary. Even poking an animal would surely be better than just looking at it: *that* would lead to anecdotalism: that was what bird-watchers did.

Yet it was also what the pioneers of ethology did. They studied natural behaviour instead of contrived behaviour, and were thus able for the first time to discern natural behaviour structures or episodes – a style of analysis helped very greatly by the comparative approach, for the occurrence of the same or similar behavioural sequences in members of related species reinforced the idea that there was a certain natural connectedness between its various terms, as if they represented the playing out of a certain instinctual programme. Then, and only then, was it possible to start to obtain significant information from the study of contrived behaviour – from the application or withholding of stimuli – for it is not informative to study variations of behaviour unless we know beforehand the norm from which the variants depart.

The form of address I chose – to trace the recent growth and transformation of ideas in four 'subjects' belonging to four levels of biological analysis – gave me no opportunity to mention some of the greatest innovations of modern scientific thought: the dynamical state of bodily constituents, for example, the perpetual flux of the material ingredients of the body. Nor have I said anything except by implication of the greatest discoveries in modern science, those which revealed the genetical functions of the nucleic acids. Yet I feel I have said enough of the growth of biology in the recent past to draw some morals, however trite.

The history of animal behaviour – in particular the sterility of the older experimental approach – illustrates the danger of doing experiments in the Baconian style; that is to say, the danger of contriving 'experiences' intended merely to enlarge our general store of empirical knowledge rather than to sustain or confute a specific hypothesis or pre-supposition. The history of embryology shows the dangers of an imagined self-sufficiency, for embryology is an inviable fragment of knowledge without genetics. (I often wonder what academics mean when they say of a certain subject that it is a 'discipline in its own right'; for what science is entire of itself?) You may think our recent history entitles us to feel pretty pleased with ourselves. Perhaps: but then we felt pretty pleased with ourselves twenty-five years ago, and in twenty-five years' time people will look back on us and wonder at our obtuseness. However, if complacency is to be deplored, so also is humility. Humility is not a state of mind conducive to the advancement of learning. No one formula will satisfy that purpose, for there is no one kind of scientist; but a certain mixture of confidence and restless dissatisfaction will be an ingredient of most formulae. Confident we may surely be, for the next twenty-five years will throw up many new ideas as profound and astonishing as any I have yet described, namely . . . but I have no time left to tell you what they are.

Two Conceptions of Science

My theme is popular misconceptions of scientific thought. I shall argue that the ideas of the educated lay public on the nature of scientific inquiry and the intellectual character of those who carry it out are in a state of dignified, but yet utter, confusion. Most of these misconceptions are harmless enough, but some are mischievous, and all help to estrange the sciences from the humanities and the so-called 'pure' sciences from the applied.

Let me begin with an example of what I have in mind. The passage that follows has been made up, but its plaintive sound is so familiar that the reader may find it hard to believe it is not a genuine quotation.

> Science is essentially a growth of organized factual knowledge . . . [true or false?] . . . and as science advances, the burden of factual information which it adds to daily is becoming well nigh insupportable. A time will surely come when the scientist must train not for the traditional three or four years, but for ten or more, if he is to equip himself to be a front-line combatant in the battle for knowledge. As things are, the scientist avoids being crushed beneath this factual burden by taking refuge in specialization, and the increase of specialization is the distinguishing mark of modern scientific growth. Because of it, scientists are becoming progressively less well able to communicate even with each other, let alone with the outside world; and we must look forward to an ever finer fragmentation of knowledge, in which each specialist will live in a tiny world of his own. St Thomas Aquinas was the last. . . .

True or false, all this? – False, I should say, in every particular. Science is no more a classified inventory of factual information than history a chronology of dates. The equation of science with *facts* and of the humane arts with *ideas* is one of the shabby genteelisms that bolster up the humanist's self-esteem. That great Platonist, Goldsworthy Lowes Dickinson, who did his best to keep science below stairs, described Aristotle as 'a man of science in the modern sense' because he was 'a careful collector and observer of an enormous range of facts.' No wonder Lowes Dickinson classified Ideas with Philosophy, Art and Love, but the sciences with – *trade*.

The ballast of factual information, so far from being just about to sink us, is growing daily less. The factual burden of a science varies inversely with its degree of maturity. As a science advances, particular facts are comprehended within, and therefore in a sense annihilated by, general statements of steadily increasing explanatory power and compass – whereupon the facts need no longer be known explicitly, i.e. spelled out and kept in mind. In all sciences we are being progressively relieved of the burden of singular instances, the tyranny of the particular. We need no longer record the fall of every apple.

Biology before Darwin was almost all facts. My friend R. B. Freeman has republished some Victorian examination questions from our oldest English school of zoology, at University College, London. The answers called for nothing more than a voluble pouring forth of factual information.[1] Certainly there is an epoch

[1] This is one (by no means the longest) of eight questions set by Professor Grant in Comparative Anatomy in February 1860:

By what special structures are bats enabled to fly through the air? and how do the galeopitheci, the pteromys, the petaurus, and petauristae support themselves in that light element? Compare the structure of the wing of the bat with that of the bird, and with that of the extinct pterodactyl; and explain the structures by which the cobra expands its neck, and the saurian dragon flies through the atmosphere. By what structures do serpents spring from the ground, and fishes and cephalopods leap on deck from the waters? and how do flying-fishes support themselves in the air? Explain the origin, the nature, the mode of construction, and the uses of the fibrous parachutes of arachnidans and larvae, and the cocoons which envelope the young; and describe the skeletal elements which support, and the muscles which move the mesoptera and the metaptera of insects. Describe the structure, the attachments, and the principal varieties of form of the legs of insects; and compare them with the hollow

in the growth of a science during which facts accumulate faster than theories can accommodate them, but biology is over the hump (though biological learned journals still outnumber learned journals of all other kinds by about three to one); and physics is far enough advanced for an eminent physicist to have assured me, with the air of one not wishing to be overheard, that the science itself was drawing to a close. . . .

The case for prolonging a scientist's formal education for many years beyond a humanist's follows naturally from the belief that scientific education is a taking on board of specialized technical knowledge. In real life, the time at which a scientist graduates is less important for scientific than for economic and psychological reasons, and for reasons to do with getting enough people through the universities in good time. The length of university schooling is far more important to those whose education ends with graduation than to those for whom education is an indefinitely continued process.

As to scientists becoming ever narrower and more specialized: the opposite is the case. One of the distinguishing marks of modern science is the disappearance of sectarian loyalties. Newly graduated biologists have wider sympathies today than they had in my day, just as ours were wider than our predecessors'. At the turn of the century an embryologist could still peer down a microscope into a little world of his own. Today he cannot hope to make head or tail of development unless he draws evidence from bacteriology, protozoology, and microbiology generally; he must know the gist of modern theories of protein synthesis and be pretty well up in genetics.

---

articulated limbs of nereides, and the tubular feet of lumbrici. How are the muscles disposed which move the solid setae of stylaria, the cutaneous invest-ment of ascaris, the tubular peduncle of pentalasmis, the wheels of rotifera, the feet of asterias, the mantle of medusae, and the tubular tentacles of actinae? How do entozoa effect the migrations necessary to their development and metamorphoses? how do the fixed polypifera and porifera distribute their progeny over the ocean? and lastly, how do the microscopic indestructible protozoa spread from lake to lake over the globe?

So it is for biologists generally. Isolationism is over; we all depend upon and sustain each other. I must not speak for specialization in the physical sciences, but feel sure that the continuous and highly successful recruitment of physicists and chemists into biology would not have been possible if they were as specialized as we are often encouraged to believe.

The thoughts I have been criticizing are thus not really thoughts at all, but thought-substitutes, declarations of the kind public people make on public occasions when they are desperately hard up for things to say.

Let me turn now to two serious but completely different conceptions of science, embodying two different valuations of scientific life and of the purpose of scientific inquiry. For dialectical reasons I have exaggerated the differences between them, and I do not suggest that anybody cleaves wholly to the one conception or wholly to the other.

According to the first conception, Science is above all else an imaginative and exploratory activity, and the scientist is a man taking part in a great intellectual adventure. Intuition is the mainspring of every advancement of learning, and *having ideas* is the scientist's highest accomplishment; the working out of ideas is an important and exacting but yet a lesser occupation. Pure science requires no justification outside itself, and its usefulness has no bearing on its valuation.

> The first man of science [said Coleridge] was he who looked into a thing, not to learn whether it could furnish him with food, or shelter, or weapons, or tools, or ornaments, or *play-withs*, but who sought to know it for the gratification of knowing. . . .

Science and poetry in its widest sense are cognate, as Shelley so rightly said. So conceived, science can flourish only in an atmosphere of complete freedom, protected from the nagging importunities of need and use, because the scientist must travel where his imagination leads him. Even if a man should spend five years getting nowhere, that might represent an honourable and perhaps even a noble endeavour. The Patrons of science – today the Research Councils and the great Foundations – should support men, not projects, and individual men rather than teams, for the history of science is for the most part a history of men of genius.

The alternative conception runs something like this: Science is above all else a critical and analytical activity; the scientist is pre-eminently a man who requires evidence before he delivers an opinion, and when it comes to evidence he is hard to please. Imagination is a catalyst merely: it can speed thought but cannot start it or give it direction; and imagination must at all times be under the censorship of a dispassionate and sceptical habit of thought. Science and poetry are antithetical, as Shelley so rightly said.[1] Scientific research is intended to enlarge human under-standing, and its usefulness is the only objective measure of the degree to which it does so; as to freedom in science, two world wars have shown us how very well science can flourish under the pressures of necessity. Patrons of science who really know their business will support projects, not people, and most of these pro-jects will be carried out by teams rather than by individuals, because modern science calls for a consortium of the talents and the day of the individual is almost done. If any scientist should spend five years getting nowhere, his ambitions should be turned in some other direction without delay.

I have made the one conception a little more romantic than it really is, and the other a little more worldly, and to restore the balance I want to express the distinction in a different and, I think, more fundamental way.

---

[1] This is not a debating point, for Shelley's writing can sustain both views. In his *Defence of Poetry* Shelley defines poetry in a 'universal' sense that compre-hends all forms of order and beauty, and includes, therefore, not merely poetry in the narrower sense, but science as well (poetry 'comprehends all science'). Earlier, however, Shelley put Reason and Imagination at opposite poles; if then, as in the second conception I outline, science is regarded as an essentially rational activity, Shelley may quite rightly be allowed to speak for the view that science and poetry are antithetical. See D. G. King-Hele in the *New Scientist* (vol. 14, pp. 352–4, 1962); and Graham Wallas, *The Art of Thought* (London, 1926).

In the romantic conception, truth takes shape in the mind of the observer: it is his imaginative grasp of *what might be true* that provides the incentive for finding out, so far as he can, what *is* true. Every advance in science is therefore the outcome of a speculative adventure, an excursion into the unknown. According to the opposite view, truth resides in nature and is to be got at only through the evidence of the senses: apprehension leads by a direct pathway to comprehension, and the scientist's task is essentially one of *discernment*. This act of discernment can be carried out according to a Method which, though imagination can help it, does not depend on the imagination: the Scientific Method will see him through.[1]

Inasmuch as these two sets of opinions contradict each other flatly in every particular, it seems hardly possible that they should both be true; but anyone who has actually done or reflected deeply upon scientific research knows that there is in fact a great deal of truth in both of them. For a scientist must indeed be freely imaginative and yet sceptical, creative and yet a critic. There is a sense in which he must be free, but another in which his thought must be very precisely regimented; there is poetry in science, but also a lot of book-keeping.

There is no paradox here: it just so happens that what are usually thought of as two alternative and indeed competing accounts of *one* process of thought are in fact accounts of the *two* successive and complementary episodes of thought that occur in every advance of scientific understanding. Unfortunately, we in England have been brought up to believe that scientific discovery turns upon the use of a method analogous to, and of the same logical stature as deduction, namely the method of *Induction* – a logically

[1] For the conception that *truth is manifest*, see the critical analysis by K. R. Popper, 'On the Sources of Knowledge and of Ignorance,' in *Conjectures and Refutations* (London, 1962). The question where Truth resides can also be put of Beauty, and answered in the same two ways, for the romantic view does not distinguish them. For the history of the idea that *beauty is manifest* (as opposed to being in the eye of the observer), see Logan Pearsall Smith, *The Romantic History of Four Words: romantic, originality, creative, genius*, S.P.E. Tract 17 (Oxford, 1924).

mechanized process of thought which, starting from simple declarations of fact arising out of the evidence of the senses, can lead us with certainty to the truth of general laws. This would be an intellectually disabling belief if anyone actually believed it, and it is one for which John Stuart Mill's methodology of science must take most of the blame. The chief weakness of Millian induction was its failure to distinguish between the acts of mind involved in discovery and in proof. It was an understandable mistake, because in the process of deduction, the paradigm of all exact and conclusive reasoning, discovery and proof may depend on the same act of mind: starting from true premises, we can derive and so 'discover' a theorem by reasoning which (if it has been carried out according to the rules) itself shows that the theorem must be true. Mill thought that his process of 'induction' could fulfil the same two functions; but, alas, mistakenly, for it is not the origin but only the *acceptance* of hypotheses that depends upon the authority of logic.

If we abandon the idea of induction and draw a clear distinction between *having an idea* and *testing it* or *trying it out* – it is as simple as that, though it can be put more grandly – then the antitheses I have been discussing fade away. Obviously 'having an idea' or framing a hypothesis is an imaginative exploit of some kind, the work of a single mind; obviously 'trying it out' must be a ruthlessly critical process to which many skills and many hands may contribute. The form taken by scientific criticism is obvious too: experimentation *is* criticism; that is, experimentation in the modern sense, according to which an experiment is an act performed to test a hypothesis, not in the old Baconian sense, in which an experiment was a contrived experience intended to enlarge our knowledge of what actually went on in nature. Bacon exhorted us, rightly too, not to speculate upon but actually to experiment with loadstone and burning glass and rubbed amber: *his* experiments answer the question 'I wonder what would happen if . . .?' Baconian experimentation is not a critical activity but a kind of creative play.

The distinction between – and the formal separateness of – the creative and the critical components of scientific thinking is shown up by logical dissection, but it is far from obvious in practice because the two work in a rapid reciprocation of guesswork and checkwork, proposal and disposal, *Conjecture and Refutation*. Though imaginative thought and criticism are equally necessary to a scientist, they are often very unequally developed in any one man. Professional judgement frowns upon extremes. The scientist who devotes his time to showing up the inadequacies of the work of others is suspected of lacking ideas of his own, and everyone soon loses patience with the man who bubbles over with ideas which he loses interest in and fails to follow up.

The general conception of science which reconciles, indeed literally joins together, the two sets of contradictory opinions I have just outlined is sometimes called the 'hypothetico-deductive' conception. For our present clear understanding of the logical structure and wider scientific implications of the hypothetico-deductive system we are of course indebted to Karl Popper's *Logik der Forschung* of 1934, translated into English as *The Logic of Scientific Discovery*.[1]

Everything I have said so far about the hypothetico-deductive system applies with exactly the same force to 'applied' science, even in its simplest and most familiar forms, as to that which is commonly called 'pure' or 'basic'. Imaginative conjecture and criticism, in that order, underlie the physician's diagnosis of his patient's ailments or the mechanic's explanation of why a car won't run. The physician may like to think himself, as Darwin did, an inductivist and a good Baconian, but with equally little reason, for Darwin was no inductivist; no more is he.

What now follows is an attempt to analyse, not the difference between basic and applied science, but the motives which have led

[1] For earlier accounts of the hypothetico-deductive scheme, see *Hypothesis and Imagination* in this volume, pp. 129-155.

people to think it highly important, and above all to make it the basis of an intellectual class-distinction.

Francis Bacon was not the first to distinguish basic from applied science, but no one before him put the matter so clearly and insistently, and the distinction as he draws it is unquestionably just. 'It is an error of special note,' said Bacon, pondering upon the many infirmities of current learning,

> that the industry bestowed upon experiments hath presently, upon the first access into the business, seized upon some designed operation; I mean sought after *Experiments of Use* and not *Experiments of Light and Discovery*.

(The image of light is a favourite of Bacon's, and the idea of kindling a light in nature.) Bacon's distinction is between research that increases our power over nature and research that increases our understanding of nature, and he is telling us that the power comes from the understanding. He felt his distinction upheld by the example of that Divine Builder who created Light only, and no *Materiate Work,* in the first day, turning to what we should nowadays presumably call Applied Science in the days following.

No one now questions Bacon's argument. Who nowadays would try to build an aeroplane without trying to master the appropriate aerodynamic theory? Sciences not yet underpinned by theory are not much more than kitchen arts. Aeronautics, and the engineering and applied sciences generally, do of course obey the Baconian ruling that what is done for use should so far as possible be done in the light of understanding. Unhappily, Bacon's distinction is not the one we now make when we differentiate between the basic and applied sciences. The notion of *purity* has somehow been superimposed upon it, and in a new usage that connotes a conscious and inexplicably self-righteous disengagement from the pressures of necessity and use. The distinction is not now between the empirically founded sciences and those whose axioms were supposedly known *a priori*; rather it is between polite and rude learning, between the laudably useless and the vulgarly applied, the free and the intellectually compromised, the poetic and the mundane.

Let me first say that all this is terribly, terribly English, or anyhow Anglo-Saxon. Making pure and applied science the basis of a class-distinction helps us to forget that it was our engineers and merchants, not the armed forces, the Civil Service and the gentry, who won for us that very grand position in the world from which we have now stepped down. It is not always easy to explain to foreigners the whole connotation of 'pure' in the context 'pure research'. They only shake their heads uneasily and wonder if it may not have something to do with cricket. They lack also our Own Very Special Blend of highmindedness and humbug in that reasoning which champions Pure Research because, while it enables the human spirit to breathe freely in the spare and serene atmosphere of the intellectual highlands, it is also a splendid long-term investment. Invest in applied science for quick returns (the spiritual message runs), but in pure science for capital appreciation.[1] And so we make a special virtue of encouraging pure research in, say, cancer institutes or institutes devoted to the study of rheumatism or the allergies – always in the hope, of course, that the various lines of research, like the lines of perspective, will converge somewhere upon a point. But there is nothing virtuous about it! We encourage pure research in these situations because we know no other way to go about it. If we knew of a direct pathway leading to the solution of the clinical problem of rheumatoid arthritis, can anyone seriously believe that we should not take it?

The more creditable part of our English reverence for pure research derives, I believe, from a certain accident of our aesthetic history. Let us concede that imaginative thought plays an impor-

[1] V. B. Wigglesworth has pointed out that in the first edition of his *Grammar of Science* (1892) Karl Pearson chose Hertzian waves, essentially radio waves, as an example of a discovery of no apparent usefulness. This is the best example I know of the apparently useless bringing in the goods (as Pearson thought it probably would).

tant part – no matter what   in discovery and invention. Now in this country the quintessential form of imaginative activity has always been poetic invention. Hereabouts, a man inspired is typically a poet inspired. Unfortunately there is no such thing as Applied Poetry – or rather, there is, but we think little of it. We look askance at poetry for the occasion, even for Royal occasions. For poetry:

> . . . is not like reasoning, a power to be exerted according to the determination of the will. A man cannot say 'I will compose poetry.' The greatest poet even cannot say it. . . .

Still less can he say that he will compose joyful or lugubrious poetry, or poetry upon a given theme.

> Poetry . . . is not subject to the control of the active powers of the mind, and its birth and recurrence have no necessary connection with the consciousness or will.

Substitute 'pure science' for 'poetry' in Shelley's manifesto, and it will help us to understand the aesthetic conspiracy which has led us to think so much more highly of pure research than of research with an acknowledged practical purpose. It was quite otherwise in the early days of the Royal Society, when there was a danger that any experiment not immediately useful would be dismissed as play.

> It is strange [said Thomas Sprat[1]] that we are not able to inculcate into the minds of many men, the necessity of that distinction of my Lord Bacon's, that there ought to be Experiments of Light, as well as of Fruit. . . . If they will persist in contemning all Experiments, except those that bring with them immediate gain, and a present Harvest, they may as well cavil at the Providence of God, that he has not made all the seasons of the year to be times of mowing, reaping, and vintage.

Sprat is not arguing for pure research in the sense in which we

---

[1] Thomas Sprat, *History of the Royal Society* (1667) p. 245. See also the *Introduction* to this volume.

should now use that term, but rather against a hasty opportunism; his formula is 'Light now, for Use hereafter'.

In countries in which poetry is not the top art form, the idea of occasional or commissioned art is commonplace and honourable, and there is correspondingly less fuss, if any, about the distinction between pure science and applied. Tapestry and statuary, stained glass, murals and portraiture, palaces, cathedrals and town halls, fire music, water music, and funeral marches – most are commissioned, and the act of being commissioned may itself light up the imagination.[1] For everyone who uses imagination knows that it can be trained and guided and deliberately stocked with things to be imaginative about. Only the irremediably romantic can believe, as Coleridge did, that artistic creation is a microcosmic version of that Divine sort of creation which can make something out of nothing, or out of a homogeneous cloud of forms or notions – and how little right *he* had to think so has been made clear by *The Road to Xanadu* and other etiological studies of Coleridge's choice of images and words. To the sober minded the 'spontaneity' of an idea signifies nothing more than our unawareness of what preceded its irruption into conscious thought.

I am labouring these obvious points in order to make it clear that poetic inspiration is not a valid guide to imaginative activity in all its forms; that there is no case for looking down on commissioned art or science or on extra-mural sources of inspiration; and that our English reverence for Pure Research, though historically understandable, and perhaps even lovable, is also slightly ridiculous.

If we study the criteria that underlie a scientist's own valuation of science, we shall certainly not find purity among them. This is a significant omission, for the scientific valuation of scientific research is remarkably uniform throughout the world.

[1] Benjamin Britten has always spoken in favour of occasional music: see his speech *On Receiving the first Aspen Award* (Faber, 1964).

Here then are some of the criteria used by scientists when judging their colleagues' discoveries and the interpretations put upon them. Foremost is their *explanatory value* – their rank in the grand hierarchy of explanations and their power to establish new pedigrees of research and reasoning. A second is their clarifying power, the degree to which they resolve what has hitherto been perplexing; a third, the feat of originality involved in the research, the surprisingness of the solution to which it led, and so on. Scientists give weight (though much less weight than mathematicians do) to the elegance of a solution and the economy of the thought and work that went into it; they give credit, too, for the difficulty of the enterprise as a whole – the size of the obstacles that had to be got over or got round before the solution was reached. But purity, as such, nowhere. Nor is usefulness, which has its own scale of valuation and its own rewards. Let usage guide us: 'How neat!' one scientist might say of another's work – or 'how ingenious!' – or 'how very illuminating!' – but never, in my hearing anyway, 'how pure!'

There is without doubt a case for uncommitted or disengaged research, but it is not self-evident, and there may turn out to be better ways of doing what pure research professes to do. For example, in institutes of basic research it is believed and hoped that something practically useful may be come upon in the course of free-ranging inquiry, whereupon research which has hitherto shed diffuse light will now come sharply into focus. This procedure works; that is, it works sometimes, and it may be the best we can do, but there's no knowing, for alternative approaches have not yet been tried out on a sufficiently large scale. Might not the converse approach be equally effective, given equal opportunity and equal talent? – to start with a concrete problem, but then to allow the research to open out in the direction of greater generality, so that the more particular and special discoveries can be made to rank as theorems derived from statements of higher explanatory value. I can see no reason why this approach, *if it were to be attempted by persons of the same ability,* should not work just as well as its more conventional alternative; in fact I believe that some great American companies are moving towards it and already have some brilliant achievements to justify their choice – e.g. the growth of a generalized communications theory out of the practical problems of sending messages by telephone. Research done in this style is always in focus, and those who carry it out, if temporarily baffled, can always retreat from the general into the particular.

If our reverence for Pure Science is a rather parochial thing, a by-product of the literary propaganda of the romantic revival; if no case can be made for it on philosophic grounds; if purity is not part of a scientist's own valuation of science; then why on earth do we think so highly of it? It is, I think, our humanist brethren who have taught us to believe that, while pure science is a genteel and even creditable activity for scientists in universities, applied science, with all its horrid connotations of trade, has no place on the campus; for only the purest of pure science can give countenance to research in the humanities – research which, though it cannot very well be described as pure, for want of anything applied to compare it with, can all too readily be described as useless. The humanist fears that if we abandon the ideal of pure knowledge, knowledge acquired for its own sake, then usefulness becomes the only measure of merit; and that if it does become so, research in the humane arts is doomed.

These fears, I have tried to explain, are groundless. Neither its purity *nor* its usefulness enter a scientist's valuation of his own research. The scientist values research by the size of its contribution to that huge, logically articulated structure of ideas which is already, though not yet half built, the most glorious accomplishment of mankind. The humanist must value his research by different but equally honourable standards, particularly by the contribution it makes, directly or indirectly, to our understanding of human nature and conduct, and human sensibility.

I have been trying to make a case for a critical study of the organization of research, by which I do *not* mean either the allocation of administrative responsibilities for research or the economics and logistics of science,[1] important though they are. I mean a study of the behavioural and intellectual structure of everything that goes into the enlargement of our knowledge and understanding of nature. I have already mentioned a few of the problems that might be on the agenda of such an investigation, and here are a few more. Are scientists a homogeneous body of people in respect of temperament, motivation, and style of thought? (Obviously not: but we talk of *the* scientist nevertheless.) Is there such a thing as a 'scientific mind'? I think not. Or *the* scientific method? Again, I think not. What exactly are the terms of a scientist's contract with the truth? This is an important question, for according to the interpretation of the scientific process which I myself think the most plausible, a scientist, so far from being a man who never knowingly departs from the truth, is always *telling stories* in a sense not so very far removed from that of the nursery euphemism – stories which might be about real life but which have to be tested very scrupulously to find out if indeed they are so.[2]

Again, and in no particular order: is it really true that a good or genuine scientist is, or should be, indifferent to matters of priority, caring only for the Advancement of Learning and nothing for who causes it to come about? How can the *frettoso* of research be combined harmoniously with the *adagio* of administration? How can the productivity of scientists be increased: is full-time research really a good thing for more than a lucky or slightly obsessional minority, and, if not, what else should a scientist do, and how should his time be parcelled out to best advantage?

These are important questions, and their answers must no longer

[1] See *The Science of Science* (London, 1964).

[2] A 'story' is more than a hypothesis: it is a theory, a hypothesis together with what follows from it and goes with it, and it has the clear connotation of completeness within its own limits. I notice that laboratory jargon follows this usage, e.g. 'Let's get So-and-so to tell his story about' something or other, an invitation which so-and-so may decline on the grounds that his work 'doesn't make a story yet' or accept because he 'thinks he's got a story.' There is a slightly depreciatory flavour about this use of 'story' because fancy has to be used to fill in the gaps and some people tend to overdo it.

be entrusted to asseveration – to 'peremptory fits of asseveration', Bacon said, when clearing the ground for his own *Great Instauration*. They will have to be thought over and argued out with some sense of urgency; and we here in England had better be quick about it, in case the wind changes and we get fixed permanently in our Anglo-Saxon attitudes to research.

Hypothesis and Imagination

'There is a mask of theory over the whole face of nature'

[1]

If an educated layman were asked to set down his understanding
of what goes on in the head when scientific discoveries are made
and of what it is about a scientist that qualifies him to make them, his
account of the matter might go something like this: a scientist is
a man who has cultivated (if indeed he was not born with) the
restless, analytical, problem-seeking, problem-solving temperament that marks his possession of a Scientific Mind. Science is an
immensely prosperous and successful enterprise – as religion is not,
nor economics (for example), nor philosophy itself – because it is
the outcome of applying a certain sure and powerful method of
discovery and proof to the investigation of natural phenomena:
The Scientific Method. The scientific method is not deductive in
character: it is a well-known fallacy to regard it as such: but it is
rigorous nevertheless, and logically conclusive. Scientific laws are
inductive in origin. An episode of scientific discovery begins with
the plain and unembroidered evidence of the senses – with innocent, unprejudiced observation, the exercise of which is one of the
scientist's most precious and distinctive faculties – and slowly
builds upon it a great mansion of natural law. Imagination kept
within bounds may ornament a scientist's thought and intuition
may bring it faster to its conclusions, but in a strictly formal
sense neither is indispensable. Yet Newton was too severe upon
hypotheses, for though there is indeed something *mere* about hypotheses, the best of them may look forward to a dignified middle
age as Theories.[1]

---

[1] *Hypotheses non sequor* runs an early draft of Newton's famous disclaimer,
which we are to translate, as Whewell did, 'I feign no hypotheses': see I.
Bernard Cohen, *Isis* **51**: 589, 1960. Newton did, of course, use and propound
hypotheses in the modern sense of that word; the unwholesome flavour which
Newton found in the word is discussed below.

A critic anxious to find fault might now raise a number of objections, among them these: (1) there is no such thing as a Scientific Mind; (2) there is no such thing as The Scientific Method; (3) the idea of naïve or innocent observation is philosophers' make-believe; (4) 'induction' in the wider sense that Mill gave it is a myth; and (5) the formulation of a natural 'law' always begins as an imaginative exploit, and without imagination scientific thought is barren. Finally (he might add) it is an unhappy usage that treats a hypothesis as an adolescent theory.

1. *There is no such thing as a Scientific Mind.* Scientists are people of very dissimilar temperaments doing different things in very different ways. Among scientists are collectors, classifiers and compulsive tidiers-up; many are detectives by temperament and many are explorers; some are artists and others artisans. There are poet-scientists and philosopher-scientists and even a few mystics. What sort of mind or temperament can all these people be supposed to have in common? *Obligative* scientists must be very rare, and most people who are in fact scientists could easily have been something else instead.

2. *There is no such thing as The Scientific Method* – as *the* scientific method, that is the point: there is no one rounded art or system of rules which stands to its subject-matter as logical syntax stands towards any particular instance of reasoning by deduction. 'An art of discovery is not possible,' wrote a former Master of Trinity; 'we can give no rules for the pursuit of truth which shall be universally and peremptorily applicable'. To many philosophers of science such an opinion must have seemed treasonable, and we can understand their unwillingness to accept a judgement that seems to put them out of business. The face-saving formula is that although there is indeed a Scientific Method, scientists observe its rules unconsciously and do not understand it in the sense of being able to put it clearly into words.

3. *The idea of naïve or innocent observation is philosophers' make-believe.* To good old British empiricists it has always seemed self-evident that the mind, uncorrupted by past experience, can passively accept the imprint of sensory information from the

outside world and work it into complex notions; that the candid acceptance of sense-data is the elementary or generative act in the advancement of learning and the foundation of everything we are truly sure of.[1] Alas, unprejudiced observation is mythical too. In all sensation we pick and choose, interpret, seek and impose order, and devise and test hypotheses about what we witness. Sense data are taken, not merely given: we *learn* to perceive.[2] 'Why can't you draw what you see?' is the immemorial cry of the teacher to the student looking down the microscope for the first time at some quite unfamiliar preparation he is called upon to draw. The teacher has forgotten, and the student himself will soon forget, that what he sees conveys no information until he knows beforehand the kind of thing he is expected to see. I cite more evidence on this point below.

4. *Induction is a myth.* In donnish conversation we are not taken aback when someone says he has 'deduced' something or has carried out a deduction; but if he were to say he had *in*duced something or other we should think him facetious if not a pompous idiot. So it is with 'Laws': Scientists do not profess to be trying to discover laws and use the word itself only in conventional contexts (Hooke's Law, Boyle's Law). (The actual usages of scientific speech are, as I shall explain below, extremely revealing.) It is indeed a myth to suppose that scientists actually carry out inductions or that a logical autopsy upon a completed episode of scientific research reveals in it anything that could be called an inductive structure of thought.

'Induction' in the wider sense that distinguishes it from perfect or merely iterative induction (see below) is a word lacking the qualities that would justify its retention in a professional vocabulary.

---

[1] The fundamental axiom of empiricism – *nihil in intellectu quod non prius in sensu* – is of course mistaken. Animals *inherit* information (for example, on how to build nests, or what to sing) in the form of a sort of chromosomal tape-recording. This instinctual knowledge is not arrived at by association of ideas, anyhow of sensory ideas received by the animal in its own lifetime.

[2] *The Organization of Behavior* by D. O. Hebb (New York, 1949), especially p. 31.

It is seldom, if ever, used in any sentence of which it is not itself the subject, and it has no agreed meaning. *Finding* a meaning for induction has been a philosophic pastime for more than a hundred years. Whewell used the word, but with some feeling in later years that he might have dropped it. 'There is really no such thing as a distinct process of induction,' said Stanley Jevons; 'all inductive reasoning is but the inverse application of deductive reasoning' – and this was what Whewell meant when he said that induction and deduction went upstairs and downstairs on the same staircase. For Samuel Neil, however, 'induction' was confined to the act of testing a scientific conjecture or presupposition, and this was also C. S. Peirce's usage ('The operation of testing a hypothesis by experiment . . . I call induction.')[1] Peirce accordingly uses the words *retroduction* or *abduction* to mean what Jevons called *induction*. Nowadays the tendency is to use 'experimentation' to stand for the acts used in testing a hypothesis, leaving 'induction' as a vague word to signify all the various ways of travelling upstream of the flow of deductive inference. (Popper, of course, is for abandoning 'induction' altogether.)

The word *experiment* has also changed its meaning. When amateurs of the history of science attribute to Bacon the advocacy of the experimental method, they are often acting under the impression that Bacon used the word as we do. But a Baconian 'experiment' had the connotation that still persists in the French *expérience* today: a Baconian experiment is a contrived experience or contrived happening as opposed to a natural experience or

[1] Here and hereafter I quote from the following works of the authors cited under heading 4 in the text:

The Philosophy of the Inductive Sciences, by William Whewell; in 2 vols., 2nd ed., London 1847 (1st ed. 1840).

The Principles of Science, by F. Stanley Jevons; 2nd ed., revised, London 1877 (1st ed. 1873).

The Art of Reasoning, by Samuel Neil: twenty articles in successive issues of the first two vols. of the *British Controversialist* (1850, 1851) of which Neil was editor: particularly No. 11, vol. 2.

Collected Papers of C. S. Peirce, eds. C. Hartshorne & P. Weiss, Harvard U.P., vol. 2 *(Elements of Logic)*, 1932; vol. 6 *(Scientific Metaphysics)*, 1935.

happening, for Bacon rightly supposed that common knowledge was not enough and that there was no relying upon luck of observation – upon 'the casual felicity of particular events'. The Philosophers of Mind took the same view: experiments were 'designed observations' intended 'to place nature in situations in which she never presents herself spontaneously to view, and to extort from her secrets over which she draws a veil to the eyes of others.'[1] Rubbing two sticks together to see what happens is an experiment in Bacon's sense; rubbing two sticks together to see if enough heat can be generated by friction to ignite them is an experiment in the modern sense. An experiment of the first kind leaves one with no answer to the question (a 'good' question: see below), 'Why on earth are you rubbing those two sticks together?'

I shall refer later to the changing connotation of 'hypothesis', a word that has grown in stature as 'induction' has declined.

The concept of induction was entrenched into scientific methodology through the formidable advocacy of John Stuart Mill. Mill, said John Venn in 1907,[2] had 'dominated the thought and study of intelligent students to an extent which many will find it hard to realize at the present day'; yet he could still take a general familiarity with Mill's views for granted, in spite of having recorded as far back as 1889 'a broadening current of dissatisfaction' on the part of physicists which had 'mostly taken the form of an ill-concealed or openly avowed contempt of the logical treatment of Induction'. It is, however, the indifference rather than the hostility of critically-minded scientists that has allowed the myth of induction to persist – combined, I believe, with the great earnestness and sincerity of Mill himself; for Mill believed, as so many good people believe today, that if only we could formulate and master The Scientific Method many of the vexed problems of modern society would vanish before its use.

[1] *Elements of the Philosophy of the Human Mind,* by Dugald Stewart, 2nd ed., London: vol. 1 1802 (1st ed. 1792), vol. 2 1816 (1st ed. 1814).

[2] *The Principles of Empirical or Inductive Logic,* by John Venn. 2nd ed., London 1907 (1st ed. 1889).

Mill's was, of course, Induction in the strong, imperfect or open-ended sense. 'Induction', said Mill ('that great mental operation')

> is a process of inference; it proceeds from the known to the unknown; and any operation involving no inference, any process in which what seems the conclusion is no wider than the premises from which it is drawn, does not fall within the meaning of the term.

That would be very well if he had not also said that induction was an exact and logically rigorous process, capable of doing for empirical reasoning what logical syntax does for the process of deduction.

> The business of inductive logic is to provide rules and models (such as the syllogism and its rules are for ratiocination) to which, if inductive arguments conform, those arguments are conclusive, and not otherwise.[1]

There seems no point in mulling over the logical errors of Mill's *System*, for they are now common knowledge – for example, his failure to distinguish between the methodologies of discovery and of proof (though Whewell had insisted on the distinction), and the circularity of his attempt to justify that 'ultimate syllogism' which had 'for its major premise the principle or axiom of the uniformity of the course of nature'. But one may yet be surprised by how little he understood the methodological functions of hypotheses, and by the hopeless ambition embodied in his belief that it was possible merely by taking thought to arrive with certainty at the truth of general statements containing more information than the sum of their known instances.

The current of informed opinion was already flowing in the other direction. The probationary character of scientific law is implicit in all of Whewell, and long before him George Campbell, in his influential and widely read *Philosophy of Rhetoric* (1776), had said of inductive generalization that there 'may be in every step,

[1] *A System of Logic,* by John Stuart Mill. 8th ed., London 1872 (1st ed. 1843).

and commonly is, less certainty than in the preceding; *but in no instance whatever can there be more'* (my italics). 'No hypothesis', said Dugald Stewart, 'can completely exclude the possibility of exceptions or limitations hitherto undiscovered.' By the latter half of the nineteenth century the point had become commonplace. 'No inductive conclusions are more than probable,' said Jevons; 'we never escape the risk of error altogether.' Venn took pains to emphasize his belief 'that no ultimate objective certainty, such as Mill for instance seemed to attribute to the results of induction, is attainable by any exercise of the human reason'. 'The conclusions of science make no pretence to being more than probable,' wrote C. S. Peirce.

The logical status of deduction and syllogistic reasoning had not been seriously in question since the days of Bacon. Syllogistic reasoning (an 'unnatural art', Campbell had called it, and others 'futile' or 'puerile') was indeed a logically conclusive process, but that was because it merely 'expands and unfolds', merely brings to light and makes explicit the information lying more or less deeply hidden in the premises out of which it flows. Deduction makes known to us only what the infirmity of our powers of reasoning has so far left concealed. The case had been well put by Archbishop Whateley,[1] and Mill accepted it; and so the peculiar and distinctive role of deduction in scientific reasoning came to be overlooked. Convinced nevertheless that Science had come upon irrefragable general truths by some process other than deduction, Mill had no alternative but to put his faith in induction – to believe in the existence of a valid inductive process even if his own account of it should prove faulty or incomplete.

What about Baconian induction – the painstaking assembly and classification of natural and elicited (experimental) facts of which Jevons said that it reduced the methodology of science to a kind of bookkeeping? By the sixth edition of the *Origin* in 1876, Darwin had convinced himself that he had been a good Baconian, but his correspondence tells a different story. Darwin's status as the culture-hero of induction – the great but deeply humble

---

[1] *Elements of Logic,* by Richard Whateley (9th ed., London 1848, 1st ed. 1826).

scientist listening attentively to Nature's lessons from her own lips – has now to be reconciled with evidence that he had the germ of the idea of natural selection before ever he had read Malthus.[1]

It is Karl Pearson whose scientific practice and theoretical professions earn him the right to be called a true Baconian. 'The classification of facts', he wrote in *The Grammar of Science*, and

> ... the recognition of their sequence and relative significance is the function of science ... let us be quite sure that whenever we come across a conclusion in a scientific work which does not flow from the classification of facts, or which is not directly stated by the author to be an assumption, then we are dealing with bad science.[2]

Poor Pearson! His punishment was to have practised what he preached, and his general theory of heredity, of genuinely inductive origin, was in principle quite erroneous.

I have given here the conventional view of Bacon's methodology, and shall return later to the claim made on his behalf by Coleridge and others that he was fully aware of the methodological value of hypotheses.

5. *The formulation of a natural law begins as an imaginative exploit and imagination is a faculty essential to the scientist's task.* Most words of the philosopher's vocabulary, including 'philosopher' itself, have changed their usages over the past few hundred years.[3] 'Hypothesis' is no exception. In a modern professional vocabulary a hypothesis is an imaginative preconception of *what might be true*

[1] *Charles Darwin,* by Gavin de Beer, p. 98 (London 1963). In his Autobiography Darwin once declared that he could not resist forming a hypothesis on every subject, and his letters to Henry Fawcett and to H. W. Bates are very revealing (*More Letters of Charles Darwin,* eds. F. Darwin and A. C. Seward, pp. 176, 195. London 1903.)

For the origin of the idea of Natural Selection, see also L. Eisely in *Daedalus* (Summer 1965), pp. 588–602.

[2] *The Grammar of Science,* by Karl Pearson, 3rd ed. London 1911 (1st ed. 1892).

[3] E.g. 'science', 'art', 'pure science', 'applied science'; 'analysis', 'synthesis'; 'experiment'; and, of course, notoriously, words like *genius, creation, enthusiasm.*

in the form of a declaration with verifiable deductive consequences. It no longer tows 'gratuitous', 'mere' or 'wild' behind it, and the pejorative usage ('evolution is a mere hypothesis', 'it is only a hypothesis that smoking causes lung cancer') is one of the outward signs of little learning. But in the days of Travellers' Tales and Marvels, when (as John Gregory contemptuously remarked[1]) philosophers were more interested in animals with two heads than in animals with one, 'hypothesis' carried very strongly the connotation of the wantonly fanciful and above all (we read it often) the gratuitous; nor was there any thought that a hypothesis need do more than explain the phenomena it was expressly formulated to explain. The element of *responsibility* that goes with the formulation of a hypothesis today was altogether lacking. Thomas Burnet's *Sacred Theory of the Earth* (1684–90) is a case in point – a romantic and absurd cosmology using the word 'hypothesis' in just the sense that Newton repudiated. 'Men of short thoughts and little meditation,' Burnet says in self-defence, 'call such theories as these, Philosophick Romances.' But, he says

> . . . there is no surer mark of a good Hypothesis, than when it doth not only hit luckily in one or two particulars but answers all that it is to be applied to, and is adequate to Nature in her whole extent.
>
> But how fully or easily soever these things may answer Nature, you will say, it may be, that all this is but an Hypothesis; that is, a kind of fiction or supposition that things were so and so at first, and by the coherence and agreement of the Effects with such a supposition, you would argue and prove that this is so. This I confess is true, this is the method, and if we would know anything in Nature further than our senses go, we can know it no otherwise than by an Hypothesis . . . and if that Hypothesis be easie and intelligible, and answers all the phaenomena . . . you have done as much as a Philosopher or as Humane reason can do.'

[1] *On the Duties and Qualifications of a Physician,* by John Gregory. New ed., London 1820 (1st ed. 1772).

Burnet's reasoning thus ends at the very point at which scientific reasoning begins. He did not seem to realize that his hypothesis about what the Earth was like before the flood and what it would be like after the Fire was but one among a virtual infinitude of hypotheses, and that he was under a moral obligation to find out if his were preferable to any other.

Burnet's preposterous speculations were expounded in prose that earned him immortality; because they were not, many philosophic romances that must have been known to Newton are now forgotten. Thomas Reid shall be allowed to sum the prevailing situation up.[1] 'It is genius,' he says, 'and not the want of it, that adulterates philosophy, and fills it with error and false theory. A creative imagination disdains the mean offices of digging for a foundation,' leaving these servile employments to scientific drudges. 'The world has been so long befooled by hypotheses in all parts of philosophy,' that we must learn 'to treat them with just contempt, as the reveries of vain and fanciful men.' Newton *could* have invented a hypothesis to account for gravitation, but 'his philosophy was of another complexion', for Newton had been 'taught by Lord Bacon to despise hypotheses as fictions of human fancy'.

Because Newton is cast as the hero of every scientific methodology of the past 200 years, philosophers who attached great importance to hypotheses felt it their duty to explain away Newton's famous and profoundly influential disavowal. Stanley Jevons was so sure that Newton had practised what is now often called the 'hypothetico-deductive' method that he was inclined to think *hypotheses non fingo* ironical. But for 200 years after Newton no one could advocate the use of hypotheses without an uneasy backward glance. Dugald Stewart said that an 'indiscriminate zeal against hypotheses' had been 'much encouraged by the strong and decided terms in which, on various occasions, they are reprobated by Newton'. 'Newton appears to have had a horrour of the term *hypothesis*,' said William Whewell. Sir John

[1] *Essays on the Intellectual Powers of Man,* by Thomas Reid, 1st ed. 1785. In *The Works of Thomas Reid,* D.D., ed. W. Hamilton, 4th ed. (London, 1854).

Herschel spoke up in favour of hypotheses.[1] Samuel Neil in 1851 deplored the 'widely prevalent prejudice in the present age against hypotheses', and Thomas Henry Huxley had felt obliged to say, 'Do not allow yourselves to be misled by the common notion that a hypothesis is untrustworthy merely because it is a hypothesis.' Even George Henry Lewes found himself unable to propound his fairly sensible views on hypotheses without much prevarication and pursing of the lips.[2]

Where does Mill stand? Modern philosophers who are for various reasons 'pro-Mill' can of course find him a devotee of hypotheses. Hypotheses, Mill will be found to say, provide 'temporary aid', even 'large temporary assistance' ('temporary' because hypotheses are the larval forms of theories); hypotheses are valuable because they suggest observations and experiments, and in this respect they are indeed indispensable. However, all this had been said before, repeatedly: some instances I shall cite later. In his less conventional utterances on hypotheses Mill betrayed that he had no deep understanding of what is now thought to be their distinctive methodological function. He feared that people who used hypotheses did so under the impression that a hypothesis must be true if the inferences drawn from it were in accordance with the facts. Later therefore he says (*System* III. 14. 6):

> It seems to be thought that an hypothesis . . . is entitled to a more favourable reception, if besides accounting for all the facts previously known, it has led to the anticipation and prediction of others which experience afterwards verified.

This, he says, is 'well calculated to impress the uninformed', but will not impress thinkers 'of any degree of sobriety'.

Mill feared the imaginative element in hypotheses: 'a hypothesis being a mere supposition, there are no other limits to hypotheses than those of the human imagination'. These are Reid's fears, preying on Mill at a time when good reasons for feeling fearful

[1] *Discourse on the Study of Natural Philosophy*. (London 1831.)

[2] *Problems of Life and Mind,* esp. pp. 296, 316-7. 4th ed., London 1883 (1st ed. 1873).

had largely disappeared. Today we think the imaginative element in science one of its chief glories. Even Karl Pearson recognized it as a motive force in *great* discoveries, but of course 'imagination must not replace reason in the deduction of relation and law from classified facts'. (The belief that great discoveries and little every-day discoveries have quite different methodological origins betrays the amateur. Whewell, the professional, insisted that the bold use of the imagination was the rule in scientific discovery, not the exception: see below.) All the same, the idea that hypotheses arose by mere conjecture, by guesswork, was thought undignified. Whewell had called good hypotheses 'happy guesses', though elsewhere, as if the occasion called for something more formal, he spoke of 'felicitous strokes of inventive talent'. But philosophers like Venn did not take to it: 'It is . . . scarcely an exaggeration of Whewell's account of the inductive process to say of it, as in fact has been said, that it simply resolves itself into making guesses.'

It is the word that is at fault, not the conception. To say that Einstein formulated a theory of relativity by guesswork is on all fours with saying that Wordsworth wrote rhymes and Mozart tuneful music. It is cheeky where something grave is called for.

II

I now turn to consider the history during the eighteenth and nineteenth centuries of some of the central ideas of the hypothetico-deductive scheme of scientific reasoning, confining myself, as hitherto, almost wholly to English and Scottish philosophers and the tradition of thought they embody. Among these ideas are:

(1) the uncertainty of all 'inductive' reasoning and the probationary status of hypotheses;

(2) the role of the hypothesis in starting inquiry and giving it direction, so confining the domain of observation to something smaller than the whole universe of observables;

(3) the asymmetry of proof and disproof: only disproof is logically conclusive;

(4) the obligation to put a hypothesis to the test.

1. I have already mentioned a number of earlier opinions on the inconclusiveness of scientific reasoning (above, pp. 136–7).

2. It is our imaginative preconception of what might be true that gives us an incentive to seek the truth and a clue to where we might find it. 'In every useful experiment,' said John Gregory, writing in 1772, 'there must be some point in view, some anticipation of a principle to be established or rejected.' Such anticipations, he went on to say, are *hypotheses*: people were suspicious of hypotheses because they did not fully understand their purpose, but without them 'there could not be useful observation, nor experiment, nor arrangement, because there would be no motive or principle in the mind to form them'. Dugald Stewart quoted passages expressing the same opinion in the writings of Boscovich, Robert Hooke, and Stephen Hales[1] – scientists all three. But on this point Coleridge sweeps everyone else aside.[2] In every advance of science, he assures us, 'a previous act and conception of the mind . . . an *initiative* is indispensably necessary', for when it comes to founding a theory on generalization,

'. . . what shall determine the mind to one point rather than another; within what limits, and from what number of individuals shall the generalization be made? *The theory must still require a prior theory for its own legitimate construction* [my italics].

Coleridge (like Stewart and later Neil) managed to convince himself of the great Francis Bacon's full awareness of the need for an 'intellectual or mental *initiative*' as the 'motive and guide of every philosophical experiment . . .

. . . namely, some well-grounded purpose, some distinct

---

[1] See Stephen Hales's Preface to his *Statical Essays,* 4th ed., London 1769 (1st ed. 1727) and a number of passages in Robert Hooke's *Posthumous Works* (London 1705). For Boscovich, see note 2, p.144.

[2] *On Method,* by Samuel Taylor Coleridge. 3rd ed., London 1849 (1st ed. 1818).

impression of the probable results, some self-consistent anticipation . . . which he affirms to be the prior *half* of the knowledge sought, *dimidium scientiae*.

The passage all three quote as evidence for this interpretation[1] is, in my reading of it, too slight to carry so great a weight of meaning.

3. Many philosophers in the older and the newer senses have spoken of the value of false hypotheses, and Stewart particularly commends the opinions of Boscovich ('the slightest of whose logical hints are entitled to particular attention'). Boscovich had said that by means of hypotheses

> . . . we are enabled to supply the defects of our *data*, and to conjecture or divine the path to truth; always ready to abandon our hypothesis, when found to involve consequences inconsistent with fact. And, indeed, in most cases, I conceive this to be the method best adapted to physics; a science in which . . . legitimate theories are generally the slow result of disappointed essays, and of errors which have led the way to their own detection.[2]

This is all right as far as it goes, but what one will not find so easily is a premonition of one of the strongest ideas in Popper's methodology, that the only act which the scientist can perform with complete logical certainty is the repudiation of what is false. It is *falsification* that has the logical stature attributed by the logical positivists to verification: 'Every experiment may be said to exist only in order to give the facts a chance of disproving the null hypothesis.'[3] The asymmetry of proof, considered as a point of logic, is of course very elementary, and it is merely slovenly or simple-minded to suppose that hypotheses are proved true if they lead to true conclusions. No logician of science has ever done so.

[1] *De Augmentis Scientiarum,* Bk. 5, Ch. 3, II; trans. Gilbert Wats. (London 1674).

[2] Dugald Stewart's translation of the footnotes on pp. 211–12 of Boscovich's *De Solis ac Lunae Defectibus* (London 1760).

[3] *The Design of Experiments* by R. A. Fisher (London 1935).

Whewell certainly realized that refutation was methodologically a strong procedure, stronger than confirmation, an opinion that comes out more clearly in his aphorisms than in the body of the text:

> (ix) The truth of tentative hypotheses must be tested by their application to facts. The discoverer must be ready, carefully to try his hypotheses in this manner . . . and to reject them if they will not bear the test.
>
> (x) The process of scientific discovery is cautious and rigorous, not by abstaining from hypotheses, but by rigorously comparing hypotheses with facts, and by resolutely rejecting all which the comparison does not confirm.

These opinions shocked Mill. Dr Whewell's system, he complained, did not recognize 'any necessity for proof':

> If, after assuming an hypothesis and carefully collating it with facts, nothing is brought to light inconsistent with it, that is, if experience does not *dis*prove it, he is content; at least until a simpler hypothesis, equally consistent with experience, presents itself.

To Mill this attitude of Whewell's betrayed 'a radical misconception of the nature of the evidence of physical truths'. No wonder Venn said that a person who read both Mill and Whewell would find it hard to believe that they were discussing the same subject! The verdict must go to Whewell, 'whose acquaintance with the processes of thought of science,' said Peirce, 'was incomparably greater than Mill's.'

4. The formulation of a hypothesis carries with it an obligation to test it as rigorously as we can command skills to do so. There was no sign of any such sense of obligation in Burnet's *Sacred Theory*: to explain the phenomena it was designed to explain was judged evidence enough. It satisfied curiosity in much the same way as a mother's desperately *ad hoc* answers satisfy the insistent questioning of a child. The child is not interested in the content of the answer: he asks as if he were under an instinctual compulsion

K

to do so, and the act of answering completes a sort of ritual of exploration. But when curiosity is satisfied it is discharged: formulation of a hypothesis may act as a deterrent rather than as a stimulus to inquiry – a danger the earlier critics of the use of hypotheses were fully aware of.

Even the more sophisticated authors of Philosophick Romances did not seem to realize that any one set of phenomena could be explained by many hypotheses other than the one they fancied. It seems a strange blindness, but I think that Dugald Stewart in a finely reasoned passage got to the bottom of it. It was a favourite conceit in eighteenth-century philosophizing – Stewart found it in Boscovich, Le Sage, D'Alembert, Gravesande and Hartley – that natural philosophy is, in David Hartley's words,[1]

> ... the art of *decyphering* the Mysteries of Nature ... so that ... every Theory which can explain all the Phaenomena, has all the same Evidence in its favour, that it is possible the Key of a Cypher can have from its explaining that Cypher.

Stewart found the analogy inept for many reasons, the chief being that whereas a cypher has one key, a unique solution, physical hypotheses seldom, if ever, 'afford the *only* way of explaining the phenomena to which they are applied'.

It is all very well to say that we are under a permanent obligation to test hypotheses, but, as Peirce said,

> ... there are some hypotheses which are of such a nature that they can never be tested at all. Whether such hypotheses ought to be entertained at all, and if so in what sense, is a serious question.

Certainly the logical positivists took the question very seriously indeed, and Popper has done so too, but I do not recollect its having been a live issue before Peirce.

III

Let me now set out the gist of the hypothetico-deductive system

[1] *Observations on Man* by David Hartley, vol. 1, pp. 15–16 (London, 1749).

as it might be formulated today. ('Gist' is the right word, for there is no question of its providing an abstract formal framework which becomes a concrete example of scientific reasoning when we fill in the blanks.) First, there is a clear distinction between the acts of mind involved in discovery and in proof. The generative or elementary act in discovery is 'having an idea' or proposing a hypothesis. Although one can put oneself in the right frame of mind for having ideas and can abet the process, the process itself is outside logic and cannot be made the subject of logical rules. Hypotheses must be tested, that is criticized. These tests take the form of finding out whether or not the deductive consequences of the hypothesis or systems of hypotheses are statements that correspond to reality. As the very least we expect of a hypothesis is that it should account for the phenomena already before us, its 'extra-mural' implications, its predictions about what is not yet known to be the case, are of special and perhaps crucial importance. If the predictions are false, the hypothesis is wrong or in need of modification; if they are true, we gain confidence in it, and can, so to speak, enter it for a higher examination; but if it is of such a kind that it cannot be falsified even in principle, then the hypothesis belongs to some realm of discourse other than Science. Certainty can be aspired to, but a 'rightness' that lies beyond the possibility of future criticism cannot be achieved by any scientific theory.

The first strongly reasoned and fully argued exposition of a hypothetico-deductive system is unquestionably Karl Popper's. Quite a large part of it had been propounded at the level of learned discourse rather than of critical analysis by William Whewell, F.R.S., Master of Trinity College, in 1840. Whewell is never heard of nowadays outside the ranks of historians of science: if one mentions his name one may be asked to spell it. But his reputation in his day was formidable. Whewell wrote upon ethics, hydrostatics, political economy, astronomy, verse composition, terminology, the Platonic dialogues, mechanics, geology and the History and Philosophy of the Inductive Sciences. He was the first *scientist*, I believe, to express a lengthy and carefully

thought out opinion on the nature of scientific discovery, and in
a sense the first scientist of any description, for he invented the
word itself.[1]

There are many inadequacies in Whewell, but the spirit is
right. No general statement, he said, not even the simplest iterative
generalization, can arise merely from the conjunction of raw data.
The mind always makes some imaginative contribution of its
own, always 'superinduces' some idea upon the bare facts. A
hypothesis is an explanatory conjecture giving one of many
possible explanations that might meet the case.

> A facility in devising hypotheses, therefore, is so far from being
> a fault in the intellectual character of a discoverer, that it is, in
> truth, a faculty indispensable to his task.

> To form hypotheses, and then to employ much labour and
> skill in refuting, if they do not succeed in establishing them, is a
> part of the usual process of inventive minds. Such a proceeding
> belongs to the rule of the genius of discovery, rather than (as
> has often been taught in modern times) to the *exception*.

Yet it is indispensably necessary for the discoverer to demand of
his hypotheses 'an agreement with facts such as will withstand the
most patient and rigid inquiry', and, if they are found wanting, to
turn them resolutely down:

> Since the discoverer has thus constantly to work his way
> onwards by means of hypotheses, false and true, it is highly
> important for him to possess talents and means for rapidly
> *testing* each supposition as it offers itself.

[1] And 'physicist' and many other useful words, including *eocene, miocene,* and
pliocene: see P. J. Wexler, *The Great Nomenclator: Whewell's contributions to
scientific terminology. Notes & Queries*, N.S. 8, p. 27, January 1961. Professor S.
Ross in *Notes and Records of the Royal Society* (**16**: 187, 1961) has recounted the
correspondence between Whewell and Michael Faraday upon what best to call
the two opposite poles of the electrolytic cell. Faraday toyed with alphode and
betaode, voltaode and galvanode, zincode and platinode, dexiode and skiaode,
oriode and occiode, eastode and westode, eisode and exode, orthode and
anthode. 'My dear Sir,' wrote Whewell, '. . . I am disposed to recommend . . .
anode and *cathode*', and so they came to be.

The hypotheses which we accept ought to explain phenomena which we have observed. But they ought to do more than this: our hypotheses ought to *foretell* phenomena which have not yet been observed [but which are] of the same kind as those which the hypothesis was invented to explain.

Whewell did not believe that a scientist acquired factual information by passive attention to the evidence of his senses; the idea of 'naïve' or 'innocent' observation (see above, p. 132) he rejected altogether: 'Facts cannot be observed as Facts except in virtue of the Conceptions which the observer himself unconsciously supplies.' The distinction between fact and theory was by no means as distinct as people were accustomed to believe: 'There is a mask of theory over the whole face of nature.' Strictly speaking, no scientific discovery can be made by accident. What Whewell has to say on Man as the Interpreter of Nature[1] is a suitable prolegomenon to Popper's famous lecture *On the sources of knowledge and of ignorance.*

The account of scientific method which became recognized as the official alternative and rival to Mill's was not Whewell's but Stanley Jevons's. Jevons is not as fresh as Whewell nor so boldly original; we may think he should have acknowledged Whewell more often than he did. Jevons gave it as his 'very deliberate opinion' that 'many of Mill's innovations in logical science . . . are entirely groundless and false'. As to Bacon, he took the 'extreme view of holding that Francis Bacon . . . had no correct notions as to the logical method by which from particular facts we educe laws of nature'. Jevons endeavoured to show that 'hypothetical anticipation of nature is an essential part of inductive inquiry', the method 'which has led to all the great triumphs of scientific research'. Even in the 'apparently passive observation of a phenomenon' our attention should be 'guided by theoretical anticipations'.

The three essential stages in the process which he continued with deliberate vagueness to call 'induction' were, in his own words,

[1] In particular see Bk. 1, Ch. 2, §§ 9, 10 (2nd ed.).

(a) Framing some hypothesis as to the character of the general law.

(b) Deducing consequences from that law.

(c) Observing whether the consequences agree with the particular facts under consideration.

Hypothesis is always employed, he says, consciously or unconsciously.

This account of the matter had come to be pretty widely agreed upon during the second half of the nineteenth century. We shall find it in Neil and Adamson[1] and very clearly in Peirce. (Venn, in spite of his reputation, I find disappointing.) There are many premonitions of the hypothetico-deductive method in the eighteenth century and even earlier, particularly in the writing of scientists. The clearest known to me is Dugald Stewart's, a point worth making because of the dismissive and totally erroneous opinion that his philosophy is simply a reproduction of his master's, Thomas Reid's, voice. In answer to Reid's rhetorical challenge to name any advance in science which had arisen by the use of a hypothetical method, Stewart thought it sufficient to mention the theory of Gravitation and the Copernican system.

Stewart believed that most discoveries in science had grown out of hypothetical reasoning:

> It is by reasoning synthetically from the hypothesis, and comparing the deductions with observation and experiment, that the cautious inquirer is gradually led, either to correct it in such a manner as to reconcile it with facts, or finally to abandon it as

[1] Robert Adamson in his article *Bacon, Francis* in the 9th ed. of the *Encyclopaedia Britannica* (Edinburgh 1875). See also Augustus de Morgan's *A Budget of Paradoxes* (London, 1872): 'Modern discoveries have not been made by large collections of facts . . . A few facts have suggested an *hypothesis*, which means a *supposition*, proper to explain them. The necessary results of this supposition are worked out, and then, and not till then, other facts are examined to see if these ulterior results are found in nature. The trial of the hypothesis is the *special object* . . . Wrong hypotheses, rightly worked from, have produced more useful results than unguided observation.'

an unfounded conjecture. Even in this latter case, an approach is made to the truth in the way of *exclusion* . . .

Stewart's own analysis of the use of those tiresome adjectives *synthetic* and *analytic* shows he is here using 'synthetically' in the sense of 'deductively'.

IV

A scientific methodology, being itself a theory about the conduct of scientific inquiry, must have grown out of an attempt to find out exactly what scientists do or ought to do. The methodology should therefore be measured against scientific practice to give us confidence in its worth. Unfortunately, this honest ambition is fraught with logical perils. If we assume for the sake of argument that the methodology is unsound, then so also will be our test of its validity. If we assume it to be sound, then there is no point in submitting it to test, for the test could not invalidate it. These difficulties I shall surmount by disregarding them entirely.

What scientists *do* has never been the subject of a scientific, that is, an ethological inquiry. It is no use looking to scientific 'papers', for they not merely conceal but actively misrepresent the reasoning that goes into the work they describe.[1] If scientific papers are to be accepted for publication, they must be written in the inductive style. The spirit of John Stuart Mill glares out of the eyes of every editor of a Learned Journal.

Nor is it much use listening to accounts of what scientists *say* they do, for their opinions vary widely enough to accommodate almost any methodological hypothesis we may care to devise. Only unstudied evidence will do – and that means listening at a keyhole. Here are some turns of speech we may hear in a biological laboratory:

[1] See Popper's *Science: Problems, Aims, Responsibilities*, in *Fed. Proc.* (Federation of American Societies for Experimental Biology), **22**: 961–72, 1963; and my own broadcast *Is the Scientific Paper a Fraud?* in *The Listener*, September 12 1963. This broadcast was followed by a correspondence (issues of September 26 and October 10) illustrating the style of thought that makes scientists treat the 'philosophy of science' with exasperated contempt.

'What gave you the idea of trying . . .?'

'I'm taking the view that the underlying mechanism is . . .'

'What happens if you assume that . . .?'

'Actually, your results can be accounted for on a quite different hypothesis.'

'It follows from what you are saying that if . . ., then . . .'

'Is that actually the case?'

'That's a good question' [i.e. a question about a true weakness, insufficiency, or ambiguity.]

'That result squared with my hypothesis.'

'So obviously that idea was out.'

'At the moment I don't see any way of eliminating that possibility.'

'My results don't make a story yet.'

'I'm still at the stage of trying to find out if there is anything to be explained.'

'Obviously a great deal more work has got to be done before . . .

'I don't seem to be getting anywhere.'

Scientific thought has already reached a pretty sophisticated professional level before it finds expression in language such as this. This is not the language of induction. It does not suggest that scientists are hunting for facts, still less that they are busy formulating 'laws'. Scientists are building explanatory structures, *telling stories* which are scrupulously tested to see if they are stories about real life.

It has been a tradition among philosophers that we should look to the physical sciences and to simple, lofty discoveries if we are to see the Scientific Method at work in its most easily intelligible form. I question this opinion. The simplicity of great discoveries is often a measure of how far they have travelled from their beginnings. Let a biologist have a turn. Here is Claude Bernard, writing just one hundred years ago:[1]

[1] *Introduction à l'Étude de la Médecine Expérimentale,* by Claude Bernard (Paris, 1865).

A hypothesis is . . . the obligatory starting point of all experimental reasoning. Without it no investigation would be possible, and one would learn nothing: one could only pile up barren observations. To experiment without a preconceived idea is to wander aimlessly.

Indeed,

Those who have condemned the use of hypotheses and preconceived ideas in the experimental method have made the mistake of confusing the contriving of the experiment with the verification of its results.

Over and over again Bernard insists that hypotheses must be of such a kind that they can be tested, that one should go out of one's way to find means of refuting them, and that 'if one proposes a hypothesis which experience cannot verify, one abandons the experimental method'. Claude Bernard is most distinctive and at his best in his insistence on the critical method, on the virtue and necessity of Doubt.

When propounding a general theory in science, the one thing one can be sure of is that, in the strict sense, such theories are mistaken. They are only partial and provisional truths which are necessary . . . to carry the investigation forward; they represent only the current state of our understanding and are bound to be modified by the growth of science . . .

This is powerful evidence, for Claude Bernard, in creating experimental physiology, did indeed put scientific medicine on a new foundation. His philosophy *worked*.

In real life the imaginative and critical acts that unite to form the hypothetico-deductive method alternate so rapidly, at least in the earlier stages of constructing a theory, that they are not spelled out in thought. The 'process of invention, trial, and acceptance or rejection of the hypothesis goes on so rapidly,' said Whewell, 'that we cannot trace it in its successive steps.' What then is the point of asking ourselves where the initiative comes from, the observation or the idea? Is it not as pointless as asking which came first, the chicken or the egg?

But this is not a pointless question: it matters terribly which came first: scientific dynasties have been overthrown by giving the wrong answer![1] It matters no less in methodology; we may collect and classify facts, we may marvel at curiosities and idly wonder what accounts for them, but the activity that is characteristically scientific begins with an explanatory conjecture which at once becomes the subject of an energetic critical analysis. It is an instance of a far more general stratagem that underlies every enlargement of general understanding and every new solution of the problem of finding our way about the world. The regulation and control of hypotheses is more usefully described as a *cybernetic* than as a logical process: the adjustment and reformulation of hypotheses through an examination of their deductive consequences is simply another setting for the ubiquitous phenomenon of negative feedback. The purely logical element in scientific discovery is a comparatively small one, and the idea of a *logic* of scientific discovery is acceptable only in an older and wider use of 'logic' than is current among formal logicians today.

The weakness of the hypothetico-deductive system, in so far as it might profess to offer a complete account of the scientific process, lies in its disclaiming any power to explain how hypotheses come into being. By 'inspiration', surely; by the 'spontaneous conjectures of instinctive reasoning', said Peirce; but what then? It has often been suggested that the act of creation is the same in the arts as it is in science:[2] certainly 'having an idea' – the formulation of a hypothesis – resembles other forms of inspirational

[1] A person who believes that the egg came first, i.e. that a new kind of egg came before a new kind of chicken, is a Mendelist, a Morganist, and to our way of thinking a regular guy, though classified until recently in Russia as a fascist villain intent upon undoing the work of the Revolution; conversely a person who believes that the chicken came first is a Lamarckist, a Michurinist and a Lysenkoist, and was at one time suspected of plotting to overthrow the constitution of the United States.

[2] See for example *Science and Human Values,* by J. Bronowski, esp. pp. 19, 27, 51. (Revised ed., New York, 1965.)

activity in the circumstances that favour it, the suddenness with which it comes about, the wholeness of the conception it embodies, and the fact that the mental events which lead up to it happen below the surface of the mind. But there, to my mind, the resemblance ends. No one questions the inspirational character of musical or poetic invention because the delight and exaltation that go with it somehow communicate themselves to others. Something *travels:* we are carried away. But science is not an art form in this sense; scientific discovery is a private event, and the delight that accompanies it, or the despair of finding it illusory, does not travel. One scientist may get great satisfaction from another's work and admire it deeply; it may give him great intellectual pleasure; but it gives him no sense of participation in the discovery, it does not carry him away, and his appreciation of it does not depend on his being carried away. If it were otherwise the inspirational origin of scientific discovery would never have been in doubt.

INDEX

Printed and bound by CPI Group (UK) Ltd, Croydon, CR0 4YY

23/10/2024

01777912-0001